MAGNETIC BUBBLES

MAGNETIC BUBBLES

T. H. O'DELL

Reader in Electronics
Department of Electrical Engineering
Imperial College, University of London

A HALSTED PRESS BOOK

JOHN WILEY & SONS
New York — Toronto

PHYSICS

First published in the United Kingdom in 1974 by
THE MACMILLAN PRESS LTD
London and Basingstoke

Published in the U.S.A. and Canada by
Halsted Press, a Division of John Wiley & Sons, Inc.,
New York

Library of Congress Cataloging in Publication Data

O'Dell, Thomas Henry.
  Magnetic bubbles.

  'A Halsted Press book.'
  1. Magnetic bubbles.
QC754.2.M34033   538'.3   74-12048
ISBN 0-470-65259-4

Printed in Great Britain by
J. W. Arrowsmith Ltd., Bristol, England

Preface

In September 1969 I spent a day at the Bell Telephone Laboratories at Murray Hill. Having always been interested in magnetism and magnetic crystals, my first request was to see the work there on magnetic bubbles. It was really good luck to meet Andrew H. Bobeck himself and spend some time with him seeing the fascinating work that he and his colleagues were doing at that time. I came back to London determined to try out some of these experiments myself and my colleague Dr E. A. D. White, Director of the Crystal Growth Laboratory at Imperial College, was just as enthusiastic. Our first experiments were done on some small bulk crystals of gallium substituted Y.I.G. but at least it was a beginning.

Since then, experimental work has led to the theoretical work which I have attempted to explain here. The book is not intended to say very much about magnetic bubble domain devices but I have tried to collect together some of the fundamental theory which is needed by anyone who takes up this interesting new field. The majority of young engineers learn very little about magnetism in their degree courses today so that the book has been written with them in mind.

I should like to thank my colleagues, Dr E. A. D. White and Dr G. E. Lane for letting me use some of their crystals. The work was helped considerably by the contact Dr White established with Dr E. A. Giess at the IBM Research Laboratories, Yorktown Heights, and this contact led to me being able to discuss a number of problems with Dr B. E. Argyle, Dr H. Chang, Dr A. P. Malozemoff and Dr J. C. Sloncewski there. They have also been kind enough to let us have copies of their many papers in advance of publication. In this country we have had the same invaluable contact with Dr R. D. Enoch and Dr M. E. Jones at the Post Office Research Laboratory. I should also like to thank Mr B. A. Boxall and Mr K. R. Papworth for their contribution to the work at Imperial College which is supported by Research Grants from the Science Research Council and the National Research and Development Corporation.

Finally, I should like to thank Dr M. E. Jones and Dr G. E. Lane for reading and commenting on the manuscript which was typed so well by Miss E. Farmer and Mrs J. Jeffery.

T. H. O'DELL
Department of Electrical Engineering
Imperial College, London

Contents

List of Principal Symbols with Their Units

M = magnetisation, amps/metre, A/m

μ_0 = $4\pi \times 10^{-7}$ henries/metre, H/m: permeability of free space

B = magnetic field, Tesla, T (10^{-4} T = 1 Gauss)

W = strip domain width, m

h = thickness of a layer of bubble domain material, m

B_0 = the externally applied bias field, T

D = the bubble domain diameter, m

E = energy, joules, J

L = inductance, henries, H

σ_w = domain wall energy density, J/m^2

F = force, newtons, N

λ = the characteristic length, m

q = a numerical constant, value 0·726

K_μ= uniaxial anisotropy energy density, J/m^3

Q = quality factor for a bubble domain material

k = Boltzmann's constant, $1·38 \times 10^{-23}$ J/°K

H = the magnetic excitation, A/m

M_p = the saturation magnetisation in an overlay bar, A/m

$\hat{M}$ = the magnetisation at the centre of an overlay bar, A/m

1 Magnetic Bubble Domains and Bubble Domain Devices

This book is about a new and very interesting entity which has been named the 'magnetic bubble domain' and how it may be used to build devices which are of great interest in electronics. In this introductory chapter, we shall begin by saying what the magnetic bubble domain is and describing some of its properties. This will immediately suggest possible applications of the magnetic bubble and the remainder of this chapter will be spent in describing some of these and showing how the bubble domain device may prove to be of great importance in the electronics of the future, particularly in the field of data processing.

The remaining chapters of the book deal with specific problems which arise when we are working with magnetic bubble domains. Chapter 2 considers the magnetostatic theory which is needed to understand how the magnetic bubble behaves in an applied field, its stability and the importance of the various physical properties of the materials which must be provided for bubble domain work. This leads to Chapter 3, where the preparation and characterisation of these materials are discussed. Chapter 4 considers dynamic problems, because we shall see in this first chapter that it is the motion of magnetic bubble domains which makes them useful in device work. In Chapter 5 we look at the problem of making bubble domains move in the controlled manner needed for device applications. The book concludes with a brief summary and some other fields of application which may be of interest.

1.1 Magnetic Domains

What is a magnetic domain? The answer to this question will be seen to explain the very important part which magnetism, or more precisely ferromagnetism, plays in electrical technology. Without the magnetic steels and alloys we should have no electric power or transport. The ceramic magnetic materials play an equally important part and are found in every radio and television receiver, in the magnetic core stores of computers and on the surface of magnetic recording tapes and discs.

The explanation for this technical importance of magnetism was perhaps best given by Kittel (1949) in the introduction to his well known review paper 'The physical theory of ferromagnetic domains'. Kittel pointed out that the essential features of ferromagnetism are characterised by the following remarkable experimental fact:

It is possible to change the overall magnetisation of a suitably prepared ferromagnetic specimen from an initial state of zero (in the absence of an applied magnetic field)

1

to a saturation value of the order of 1000 gauss, by the application of a field whose strength may be of the order of 0·01 Oersted.

Units are traditionally confused in magnetism and the above quotation would be more immediately dramatic if we said that the magnetic field inside the ferromagnet changed from zero to 10^3 when the externally applied field changed from zero to 0·01. The material, it would appear, has amplified the magnetic effect by a factor of 10^5. This would be the situation in a long thin sample of material, or a sample in the form of a closed ring.

How does this remarkable behaviour come about? The answer is illustrated by figure 1.1 which shows a rectangular bar of magnetic material inside a coil through which current may be passed. When there is no current flowing in the coil, the magnetic material is in the demagnetised state. This really means that small volumes inside the bar, called ferromagnetic domains, will be found fully magnetised in certain directions but, taken overall, just as much of the bar will be found magnetised in any one direction as another so that the net magnetisation adds up to zero.

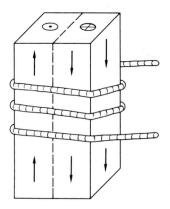

Figure 1.1. The demagnetised state; equal volumes of material are magnetised both up and down.

Things have been made particularly simple in figure 1.1 by looking at what we would call a uniaxial ferromagnet, that is, one which can only be magnetised along one particular direction. A further simplification has then been made by supposing that one half of the bar shown in figure 1.1 is one single magnetic domain which is magnetised upwards while the other half is a single domain magnetised downwards. The net result is zero overall magnetisation.

Now consider what happens when a small current is passed through the coil, as shown in figure 1.2. Experimentally, we find that the very small magnetic field produced by this current may be sufficient to move the boundary between the two domains a considerable distance and so produce

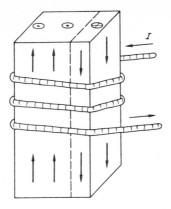

Figure 1.2. A small magnetising current may produce a large change in the net magnetisation.

a large net magnetisation. The magnetic effects which we used in technology rely upon behaviour of this kind—the existence of ferromagnetic domains and the movement of the domain walls which separate them. The early development of magnetic materials might be summarised by saying that there were two main lines of effort—one group of people were concentrating on making the domain walls move more easily (high permeability, low loss materials) while the other group were trying to stop them moving completely (materials for permanent magnets).

1.2   Single Crystals of Magnetic Material

The familiar magnetic alloys and ceramics are polycrystalline; a mass of microcrystals, usually a few microns in diameter, usually orientated randomly. The true situation inside a rectangular bar of magnetic steel, for example, which finds itself in the situations shown in figures 1.1 and 1.2, would be that the bar would be divided up into a very large number of magnetic domains. It might be found that every microcrystal in the bar contained several magnetic domains, all magnetised along different, but corresponding, crystallographic directions, or it could be that each ferromagnetic domain contained several microcrystals, and then the easy directions of magnetisation would be decided by the internal stresses of the material. In either case, it is immediately obvious that these polycrystalline materials are going to show very complex magnetic behaviour because the movement of a very large number of domain walls will be involved whenever the state of magnetisation is changed by applying external magnetic fields.

If it is possible to prepare a single crystal specimen of magnetic material, the whole problem of the magnetic domains becomes far more straightforward. A single crystal is completely homogeneous on an atomic scale and the magnetic domains will be found magnetised along particular crystallographic directions. In the three most well-known magnetic elements, iron, nickel and cobalt, for example, we find that the direction of magnetisation is normally along the cube edge, the cube diagonal and the hexagonal axis, respectively. The early experiments on single crystals of magnetic material were done on these metal crystals quite early in the twentieth century and are reviewed by Bozorth (1961). Later work on single crystals of metal alloys is reviewed by Kittel (1949). The metal crystals are, as a rule, very difficult to grow because the crystal structure just below the melting point is usually quite different to the one which is found when the metal is cooled to room temperature. For this reason, and because they are very easily destroyed by the slightest mechanical stress, the metal single crystals have not found any great application technically. In addition, metals are electrical conductors which is not conducive to good high frequency performance in a magnetic material.

A very different situation applies when we look at the magnetic oxide materials which began to make a serious impact upon technology around 1947 when the work done by Snoek during the second world war on the ferrites was published (Snoek, 1947). Single crystals of the ferrite materials were soon grown (Galt *et al.*, 1950) and this was followed by the growth of single crystals of the other oxide materials—the orthoferrites (Remeika, 1956) and then the magnetoplumbites and garnets (Nielson and Dearborn, 1958).

The growth technique for single crystals of the magnetic oxides usually involves making a solution of the magnetic material at very high temperature in a flux, such as lead oxide, and allowing this solution to cool very slowly. If conditions are right, a few crystals will nucleate when the solution becomes supersaturated and these will grow as cooling proceeds. The resulting crystals are very different from the single crystals of the metals and alloys in that the oxides are very hard refractory materials which, once grown as single crystals, can be cut and polished to any desired shape. In addition, the oxide materials are nearly all excellent insulators so that they are useful as magnetic materials at very high frequencies. Single crystals of the garnets, for example, were soon used in a multitude of microwave devices.

1.3   Magnetic Domains in Single Crystals

1.3.1   *Observing domains*

A remarkable property of the oxide magnetic materials is that they can be made reasonably transparent. This follows because of their insulating properties and because the optically excitable atoms in their structure may be quite dilute. When a magnetic material is transparent it is possible to

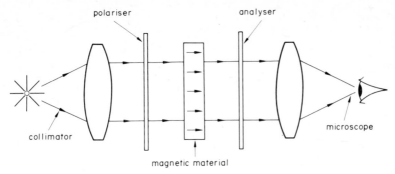

Figure 1.3. The Faraday effect. Polarised light has its plane of polarisation rotated when it passes through a magnetic material—clockwise or anticlockwise depending upon whether the direction of propagation is the same as the direction of magnetisation or in the opposite direction.

actually see the magnetic domains inside it by means of the Faraday effect, which is illustrated in figure 1.3.

Let us again consider a uniaxial ferromagnetic material, as we did in figures 1.1 and 1.2, but, this time, let us suppose that it is in the form of a thin single crystal sheet, as shown in figure 1.4, cut so that the easy direction of magnetisation is normal to the surface of the sheet. If the material is strongly anisotropic, which means that it is very difficult for the magnetisation to deviate from the easy direction, we shall find that the sheet is split up into magnetic domains, as shown in figure 1.4(a), with their directions of magnetisation alternately up and down, producing zero net magnetisation overall.

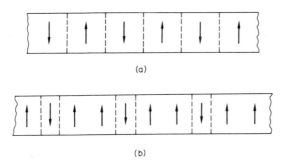

Figure 1.4. A thin single crystal sheet with its easy direction of magnetisation normal to its plane. The demagnetised state is shown in (a) and the effect of an applied field, upwards, is shown in (b).

These magnetic domains can now be observed if we mount the sheet, as shown in figure 1.3, in a beam of polarised light. Because of the Faraday effect, the plane of polarisation of the light will be rotated in different directions for the two different directions of magnetisation which are found in the sheet. The angle of the analyser shown in figure 1.3 can be adjusted to extinguish the light which has passed through one kind of domain, the ones magnetised away from the observer, for example, so that these domains appear dark. Domains magnetised towards the observer will then appear light.

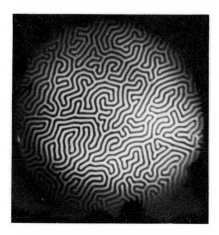

Figure 1.5. Faraday rotation micrograph showing the magnetic domains in a $10\,\mu$ thick epitaxial garnet layer.

Figure 1.5 shows such a picture, or Faraday rotation micrograph, taken of a thin sheet of magnetic garnet. Here we see equal areas of light and dark serpentine magnetic domains, showing that a section through the sheet would, in fact, show the state of magnetisation shown in figure 1.4(a). Because this particular garnet is very isotropic in its in-plane properties, the magnetic domains wind around one another in an entirely random manner. This pattern is the one which we normally observe in such single crystal layers when there is no applied field. It is the demagnetised state, which was idealised in figure 1.1. Equal volumes of the sample are magnetised towards us and away from us.

1.3.2   *The effect of an applied field*

We now consider the effect of an externally applied magnetic field on the thin sheet shown in figure 1.4(a). This field, B_0, will be continuous right through the sheet, because of the fundamental relationship div $\mathbf{B} = 0$, and will increase the energy density of the domains which are opposed to B_0, that is, the domains magnetised downwards. Similarly, the energy density of the domains which are magnetised in the same direction as B_0, upwards, will be decreased. It follows that there will now be a force upon the domain walls which will cause these to move in such a way that the high energy domains contract and the low energy domains expand. This will produce a net magnetisation upwards, as shown in figure 1.4(b).

The effect upon the Faraday rotation micrograph is shown in figure 1.6. The domains which appeared dark in figure 1.5 have become thinner as a result of the field applied in a direction towards the observer.

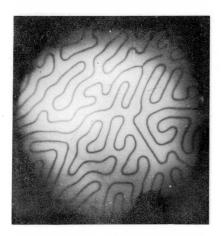

Figure 1.6. The effect of a bias field upon the domains shown in Figure 1.5. The field of view is 0·5 mm in diameter.

1.3.3 *The magnetic bubble domain*

It is from figure 1.6 that we may introduce the entity which is the subject of this book—the magnetic bubble domain. Figure 1.6 shows the situation which comes about when a constant field, B_0, is applied in a direction normal to the surface of the thin sheet. This field squeezes the dark domains, shown in figure 1.5, into the narrow strip domains of figure 1.6. Suppose we now add a pulsed magnetic field to B_0. At the instant the pulsed field is applied, the strips shown in figure 1.6 begin to contract still further and, if the pulse were sufficiently intense and were to last a sufficient length of time, these strips would collapse completely and leave the thin sheet of magnetic material uniformly magnetised, or saturated, in the direction of the applied field.

It turns out that this contraction cannot occur absolutely uniformly but happens in the wave-like manner shown in figure 1.7. Before collapsing completely, the thin strip domains of figure 1.6 actually break up into small parts which then contract symmetrically, figure 1.7(e). If, at this point, the pulsed field is switched off, we find ourselves left with these small isolated domains, which expand slightly and turn out to be stable entities in the constant bias field, B_0. The domains shown in figure 1.7(e) are small cylindrical domains. They are the magnetic bubble domains which we are going to study here.

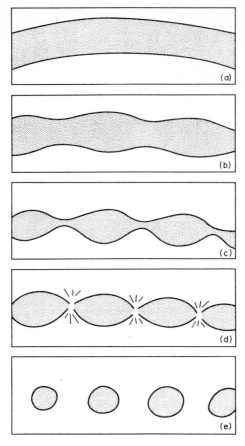

Figure 1.7. Showing the contraction of a thin strip domain, (a), occurring in a wave-like manner, (b) to (c), and finally breaking up, (d), to form bubble domains, (e).

Figure 1.8 shows this effect actually happening. The narrow strips of figure 1.6 have been subjected to just one field pulse and we can see that some of them have broken up into magnetic bubble domains. Some correlation between the original strips of figure 1.6 and the strings of bubble domains in figure 1.8 can be seen, although this is not necessarily to be expected because the bubbles, once created, can move quite freely around and will do so because of the weak dipole interactions they have with one another and with the remaining strip domains. This is further illustrated in figure 1.9 which shows the result of applying several more pulses, after the photograph for figure 1.8 had been taken. It can be seen that all the strips have now been broken up into bubble domains, in figure 1.9, and that these have arranged themselves into a fair hexagonal lattice because of their mutual repulsion.

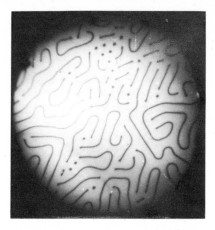

Figure 1.8. A pulsed magnetic field is applied to the domains shown in Figure 1.6. Magnetic bubble domains are generated.

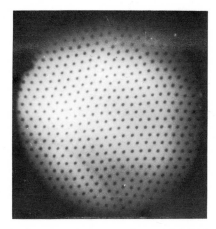

Figure 1.9. Repeated pulses generate an array of magnetic bubbles. In this particular garnet, these bubbles are $\approx 6\,\mu$m in diameter.

1.3.4 Bubble stability

The magnetic bubble domain shown in figures 1.8 and 1.9 is a small cylindrical magnetic domain passing right through a single crystal layer of uniaxial magnetic material. In Chapter 2 we shall see how it is that the bubble domain is held in stable equilibrium by the squeezing force of the applied field being opposed by the internal magnetic pressure which is produced by the magnetisation within the bubble itself. This is illustrated in figure 1.10. The circular shape of the bubble domain is maintained because of a third force; the magnetic surface tension of the wall which surrounds the bubble domain.

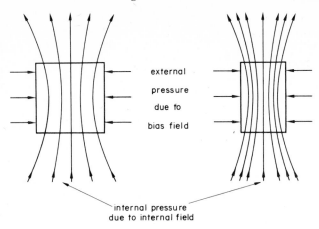

external

pressure

due to

bias field

internal pressure
due to internal field

Figure 1.10. A contraction of the bubble domain, produced by an increase in the bias field, will be opposed by an increase in the internal field of the domain itself.

Magnetic bubble domains are thus well named. When viewed in the microscope, they can be made to appear surprisingly like small soap bubbles simply by moving a small magnet close to the end of the objective lens. In such a non-uniform field, the bubble domains are seen to 'float' slowly towards regions where the applied field is weaker. In moving, they are seen to avoid one another, and, this again, seems to be a natural kind of behaviour for 'bubbles' because the soap bubbles we can blow are usually electrically charged and avoid contact with one another for this reason. The magnetic bubbles, in contrast, are dipoles and avoid contact with one another because all their dipole moments are constrained to lie in the same direction, the cylinder axis. We view the cylindrical magnetic bubbles end on so that the similarity between them and spherical soap bubbles is quite striking.

1.3.5 *Bubble domain mobility*

It is precisely this ability to move, or mobility, of bubble domains which suggests their application in data processing devices. The device idea may be summarized as follows—let us suppose we have worked out some way of moving magnetic bubble domains along some predetermined track which is marked out on the surface of a thin layer of bubble domain material. Every bubble domain is supposed to be moving at the same velocity. We may now represent a binary number by means of a string of bubble domains, the 'ones' being represented by bubbles and the 'zeroes' being represented by the absence of bubbles.

In such circumstances we have a situation very similar to a magnetic recording tape carrying binary data. There is one striking difference, however, and that is that the magnetic pattern is moving within the stationary magnetic material. A closed track, marked out on the surface of a layer of bubble domain material could, in principle, be used as a data store in the same way as a closed loop of magnetic tape. Such a bubble domain device would involve no mechanical motion, however, and because the bubble domains are only a few microns in diameter and may have velocities, we shall see later, of several metres per second, a data rate of megabits per second would be possible. This data rate applies to just one bubble domain track and the really important point about bubble domain devices is that we are talking about processes occurring in two dimensions. This opens up a completely new range of possibilities. For example, a parallel to serial converter in bubble domain technology seems almost trivial—it would involve a single track which would have n tracks joining it on the side. An n bit word in parallel form could then pass down the n side tracks and be transferred into the single track where it could then be made to propagate as a time serial representation of the parallel input. The same hardware is made a serial to parallel converter by reversing the input and output.

It is this two dimensional manipulation of data, represented by means of bubble domains, which makes the bubble domain device of considerable potential importance. Logic and storage become intimately connected. Binary numbers stored as bubble domain strings may be manipulated to lie physically side by side within the layer of bubble domain material so that they can interact. A bubble domain in one track can be made to switch the progress of a bubble domain in a neighbouring track from one branch to another. This will be considered in more detail in §1.4.5.

Ideas of this kind were suggested long before the magnetic bubble domain was thought of and had even been realised to a certain extent in the same field of magnetics. Broadbent (1960), for example, had suggested very similar ideas using thin ferromagnetic films in which the magnetisation lay in the plane of the film. A magnetic pattern could be made to propagate across the surface of such a film and Broadbent constructed a shift register using this technique, which worked at 1 MHz, and also pointed out the two dimensional possibilities of devices of this kind. Even earlier, Bobeck and Fischer (1959) explored the possibilities of moving magnetic domains, but these were in only one dimension being along the length of a magnetic wire.

The real difference between these early ideas and the magnetic bubble was that Bobeck (1967) realised that, for a magnetic domain to be truly mobile in two dimensions, the magnetisation of the domain must be perpendicular to the plane of the layer in which it moves. There is then no preferred direction in the plane of the layer and the location of the bubble domains is undefined until we deliberately introduce some kind of guiding track along which they can be made to move. It is an interesting historical point that the first photograph published of the magnetic bubble domain was probably the one given by Sherwood, Remeika and Williams (1959) showing a bubble domain in $YFeO_3$. It was not until 1967, however, that Bobeck (1967) published his now famous paper 'Properties and device applications of magnetic domains in orthoferrites' in which the device potential of bubble domains was first described.

1.4   Bubble Domain Devices

The remaining chapters of this book will deal with the theoretical problems which are involved in the bubble domain device and the treatment will be made as general as possible. The technical details of working devices are subject to rapid development and are described in a rapidly growing literature. The remainder of this chapter is therefore intended to serve as an introduction to the bubble domain device by illustrating some of the techniques which are used. These examples have been selected from the current literature not because they are necessarily the best examples of bubble domain device work but because they show up the problems for which we need the theory given in the later chapters.

1.4.1   *Bubble propagation*

The first requirement of a device which exploits the magnetic bubble is to have some kind of propagation track. This could be said to be analogous to the transmission lines of conventional electronic systems in that the bubble domain propagation track carries the signal, now a bubble domain signal, from one part of the system, or device, to another. This signal is a truly magnetic signal and is intrinsically a binary signal, ones being represented by bubbles and zeroes being represented by the absence of bubbles.

The bubble domains shown in figures 1.8 and 1.9 are supported by the externally applied bias field, which is in the opposite direction to the magnetisation within the domain. This bias field produces an external pressure around the bubble domain, which tends to make it contract, and this pressure is opposed by the internal magnetic pressure of the field within the domain, as illustrated by figure 1.10.

It follows that bubble domain propagation can be accomplished by making the bias field vary over the surface of the magnetic layer which is supporting the bubble domains. For example, suppose that the bias field is a periodic function of x, as shown in figure 1.11. A bubble domain in such a periodic field would experience a higher external pressure on one side compared to the other, unless it found itself symmetrically placed about a minimum in the periodic field. Such a field minimum is thus a stable bubble location as figure 1.11 shows. Bubbles to either side of this stable location will experience a net force which will tend to move them into the stable location.

If the periodic field, shown in figure 1.11, is now made a travelling magnetic wave, the bubble domains will be carried with the wave in locations very close to the field minima. This is shown in figure 1.12 where a bubble domain is shown moving in a field $B_z = B_0 + \hat{b}_0 \sin 2\pi(x/\lambda - ft)$, B_0 being

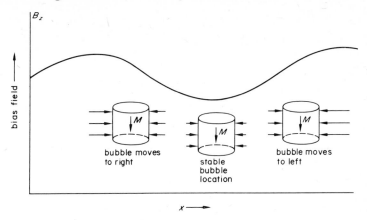

Figure 1.11. A bubble domain in a bias field, B_z, which varies periodically along x will take up a position at a field minimum.

the constant component of the field and $\hat{b}_0$ the amplitude of the travelling wave component.

In practice a propagation track would be required which could be used to guide a string of bubble domains along any desired path over the surface of the magnetic layer. A practical realisation of this is shown in figure 1.13 in which the travelling wave part of the field is produced by a conductor pattern, driven by a two phase current supply, the constant part of the bias field would be produced by some external magnet and the stability of the propagation track is ensured by the small permalloy dots shown in figure 1.13. These permalloy dots also provide stable locations for the bubble domains in the absence of any drive currents. This is a very desirable property of any bubble domain device propagation track because it implies that a string of bubble domains, representing binary coded data, will preserve its integrity when the device is not activated.

Propagation tracks involving conductor patterns are certainly the easiest to design but they are very difficult to make for bubble domains which are only a few microns in diameter. A further disadvantage is that, even when we can make such fine and narrow conductor patterns, the currents which will be required in these conductors may be embarrassingly high.

For this reason, there is considerable interest in making propagation tracks which involve 'field access'. This means that a very large number of propagation tracks on a bubble domain device are all controlled and made to work in synchronism by applying one external magnetic field to the whole device. These propagation tracks combine very naturally with some of the other operations, such as steering and duplication, which we need to do within bubble domain devices and can be realised by using patterns of magnetic material which are printed, by photolithography or some similar

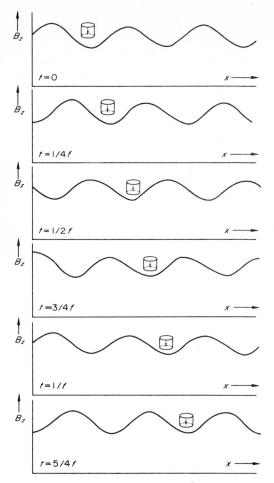

Figure 1.12. A travelling magnetic wave, $B_z = B_0 + \hat{b}_0 \sin 2\pi(x/\lambda - ft)$, carries the magnetic bubble along with it.

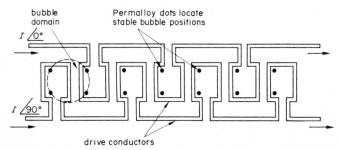

Figure 1.13. A bubble domain propagation track which uses a conducting pattern, upon the surface of the magnetic material, driven by a two phase supply (Bobeck *et al.*, 1969).

technique, on, or very close to, the surface of the magnetic layer which supports the bubble domains.

These magnetic overlays are the subject of Chapter 5 and operate rather subtly. At this introductory stage we shall avoid any discussion of the theory and take a qualitative look at their properties and possibilities.

1.4.2 *Bubble propagation with magnetic overlay patterns*

As figure 1.12 shows, bubble domain propagation requires the production of a travelling magnetic wave, $\hat{b}_0 \sin 2\pi(x/\lambda - ft)$, which is superimposed upon the constant bias field B_0. Figure 1.14 shows the principle which can be used to produce such a travelling magnetic wave by means of a magnetic

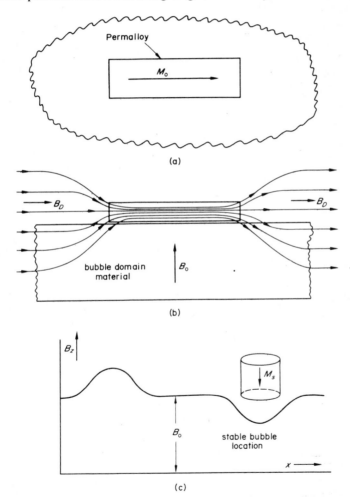

Figure 1.14. An in-plane field B_D can modify the bias field, B_z, by acting through a permalloy overlay.

16   *Magnetic Bubbles*

overlay which is printed on top of the magnetic layer supporting the bubble domains.

In figure 1.14, the single element of a magnetic overlay of permalloy is shown which is magnetised to saturation, M_0, along its length. This state of magnetisation is produced by means of an external field, B_D, the drive field, as shown in figure 1.14(b). In the absence of any bubble domains, the field B_D will be concentrated within the permalloy overlay and a component of B_D will be produced which aids the bias field, B_0, at one end of the permalloy and opposes it at the other end. This will produce a variation in the normal component of the field as shown in figure 1.14(c). It follows that a stable bubble location will be produced at one end of the permalloy strip.

We shall see, in Chapter 5, that the situation is complicated by the interaction between the bubble domain and the permalloy overlay. This interaction, in fact, aids the formation of the stable bubble location. It is sufficient at this introductory stage to ignore this interaction and see how a travelling magnetic wave can be produced by *rotating* the drive field, B_D, and this is shown in figure 1.15.

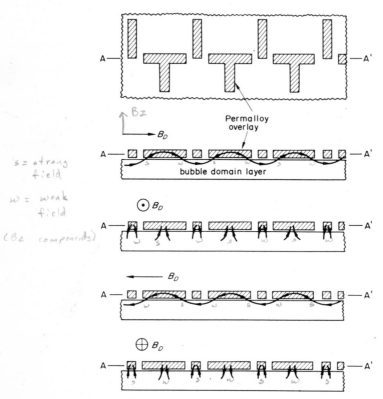

Figure 1.15. An in-plane field, B_D, acting through a pattern of permalloy **T** and **I** bars. As B_D is rotated a travelling magnetic wave is produced along A–A'.

Figure 1.15 shows a well known bubble domain propagation track; the **T** and **I** bar pattern first described by Perneski (1969). In this pattern we are exploiting the fact that a rectangular overlay of permalloy will produce a much larger external field when its magnetisation lies along its length, as it does in figure 1.14, than when it lies across the shorter dimension. This is due to the much larger demagnetising effect found in a short broad magnet compared to a long thin one. Figure 1.15 shows that, as the drive field rotates, the minimum of the field normal to the surface of the bubble domain layer is made to move from left to right simply because of the way in which the long dimensions of the **T** and **I** bars are arranged relative to one another. The direction of propagation can be reversed, for the same sense of external field rotation, by simply inverting the **TI** structure shown in figure 1.15 to a **⅃I** structure. It follows that a very long propagation track may be laid over the surface of a magnetic bubble domain layer by using alternate lines, **TI**, **⅃I** joined by a corner propagation pattern working on the same principle.

1.4.3   *Bubble domain generation*

Having obtained a defined propagation track for the magnetic bubble domains we now need some method of generating bubbles and launching them in to the propagation track. When we first introduced the bubble domain here, in figures 1.8 and 1.9, generation was accomplished by means of a pulsed magnetic field. Generation in a device may be done in this way (Bobeck *et al.* 1969) but the method does call for rather large pulsed currents to flow in very fine metallisation patterns. To avoid this, generation is usually accomplished by some technique which is compatible with the magnetic overlay patterns, discussed above, and the rotating in-plane field.

Figure 1.16 shows a very simple generator of this type which is due to Perneski (1969). The bubble generator part of this pattern consists of a circular disc of permalloy, perhaps five times the diameter of an isolated bubble domain in the material being used, and a source domain has been previously nucleated underneath this disc. The source domain is held at the edge of this disc, no matter what direction the magnetisation, M_0, of the disc, because there will always be a field minimum at some point around its periphery. The source domain is, of course, very considerably distorted from the bubble shape because of this large disc of permalloy. As the in-plane drive field rotates, it can be seen that bubble domains are launched on to the propagation track because of the rectangular tab which has been joined on to the disc. Every time the drive field completes one revolution, one bubble is launched on to the track and all the bubbles, which are already on the track, move along one pattern period. As shown in figure 1.16, this generator generates a continuous stream of 'ones' in the binary code.

A useful generator must be modulated so that its signal can carry information. We should like to be able to modulate a generator in two very

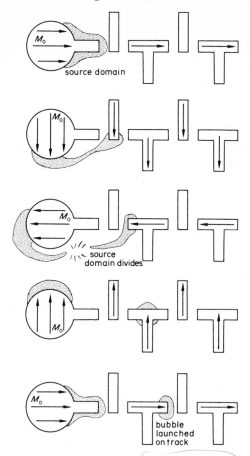

Figure 1.16. The bubble domain generator due to Perneski (1969), the in-plane magnetisation of the permalloy overlay, M_0, is shown rotating as the drive field is rotated.

distinct ways—

(a) Modulation by means of an external electrical signal.

(b) Modulation by means of a second bubble domain signal propagating within the same device.

The second possibility emphasises the very special nature of the bubble domain device in that bubble domain signals can interact with one another; a point which was discussed at the beginning of this chapter. Such a modulation of the bubble domain generator is thus a bubble interaction problem and will be discussed in § 1.4.5.

Figure 1.17 shows a method of modulating the generator shown in figure 1.16 with an external electrical signal. It can be seen that this device does involve an electrical conductor pattern, a feature which we have re-

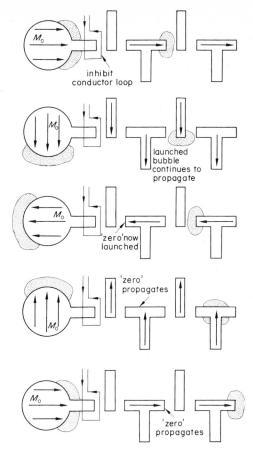

Figure 1.17. The bubble domain generator of figure 1.16 with an inhibit conductor loop.

peatedly tried to avoid. In this case, however, the current required in order to inhibit the generation process will be quite small because the conductor pattern field is associated with the permalloy overlay. The modulated bubble domain generator shown in figure 1.17 simply works by arranging that the field minimum at the end of the rectangular tab, shown by the presence of the domain at this point in the first and last parts of figure 1.16, is made a field maximum. The domain then avoids the rectangular tab and does not transfer to the propagation track (Chang *et al.* 1972).

It follows that a pulsed binary signal, passed through an inverter and then applied to the inhibit conductor loop, will be converted from a serial electrical binary coded signal into a serial bubble domain binary coded signal. The input signal and the rotating drive field must, of course, be in synchronism. Figures 1.16 and 1.17 can be looked at as one sequence showing the generation of the signal '01' in bubble form. This would be a time serial '01'. A parallel representation would be more normal in a data processing system and this would simply involve a number of generators and their associated propagation tracks, all on the same bubble domain layer. The electrical signal would then enter the bubble domain device in parallel, along n lines connected to n inhibit loops of the kind shown in figure 1.17. Parallel representation, however, would usually require a more sophisticated form of addressing than this and we shall consider this problem again in § 1.4.5.

1.4.4   *Bubble domain detection*

Given a bubble domain generator, of the kind shown in figure 1.17 which can be modulated with an incoming electrical signal, and a propagation track, of the kind shown in figure 1.15, it would be possible to build a simple data store. This would be done by making the propagation track a closed loop so that the data launched on to the track would circulate continuously.

Stored information is only useful when it can be read out at some later stage so that the next essential element of our bubble domain device must be some kind of detector. In the case of a closed loop data store, the detector would, ideally, be a non-destructive type of sensing element which could be placed close to the propagation track at some point and read out the data, as it passed this point in the form of magnetic bubble domains, by producing an electrical pulse for each passing bubble.

Successful detectors of this kind have been made which use magneto-resistive elements (Almasi *et al.* 1972) and Hall effect sensors (Strauss and Smith 1970) to detect the stray magnetic field of the bubble domain as it passes the sensor. The magneto-resistive sensor has been particularly successful and will be considered in some detail in Chapter 5, § 3. Optical techniques have also been suggested (Almasi 1971). An alternative idea is to duplicate the bubble domain signal as it circulates around the closed storage loop, lead it off by means of a branch propagation track and then detect this signal destructively by means of an inductive sensor (Chang *et al.* 1972). A destructive technique has the advantage that all the energy contained in the bubble domain can be coupled into the sensor whereas the non-destructive techniques, the magneto-resistive or Hall effect sensors, can only couple with a small fraction of this available energy.

1.4.5   *Bubble domain steering, duplication and logic*

Bubble domain generation, propagation and detection have been discussed above and these are the most important features of the new bubble domain technology. By these means we are able to convert binary electrical signals, propagating along electrical conductors, into binary magnetic signals, propagating as bubble domains along magnetic overlay tracks, and then convert these magnetic signals back into electrical signals. From a data storage point of view, this is very similar to the delay line storage which was used in very early digital computers. With bubble domains, however, we have the remarkable possibility of varying the propagation velocity right down to zero and of actually reversing it.

A much more advanced level of device can be made, however, if we can work out ways of switching bubble domain signals from one track to another, duplicate signals, which is the equivalent of amplification, and finally make these bubble domain signals interact with one another so that new signals can be generated which bear some previously defined logical relationship to two or more input bubble domain signals. These processes must be done within the bubble domain medium itself, not by detecting the bubble domain signal and then processing it by the established semiconductor hardware.

For the purposes of introduction, figures 1.18, 1.19 and 1.20 show some of the possibilities which can be realised using the kind of magnetic overlay pattern which has been considered so far. In figure 1.18 we show the simplest kind of bubble steering, or switching, where a bubble domain signal is switched between two possible propagation tracks by means of an external electrical signal. Such a steering circuit would be needed to construct an addressing system for a number of closed loop propagation tracks, such as those found in a bubble domain mass memory (Bosch *et al.* 1973).

Figure 1.18(a) shows bubble domain propagation along a straight-forward **TI** propagation track which has a branch connected to it at the bottom of the second **T**. This **T** has a conductor loop placed over it and when the loop is not carrying any current, as shown in figure 1.18(a), the bubbles propagate along the upper track as the drive field rotates and no bubbles enter the lower track.

In figure 1.18(b), the steering conductor loop current is switched on just at the time a bubble domain enters the second **T** in the upper track. After the next quarter cycle of the drive field, this bubble finds itself held at the head of the **T** by both the drive field and the field of the steering loop so that, when the next quarter cycle takes place, the bubble domain does not move on but remains under the steering loop. The next quarter cycle shown at the bottom of figure 1.18(b) causes the bubble domain to move right down the **T** and enter the lower track. This happens because the field maximum now produced at the head of the **T** by the drive field cancels out the field minimum produced by the steering loop so that the bubble must move down the **T** and remain at the field minimum which the drive field has produced at the bottom of the **T**. Comparing figures 1.18(a) and (b) shows that a 'zero' is now propagating in the upper track compared to the 'one' which was shown in figure 1.18(a). Similarly a 'one' is launched into the lower track of figure 1.18(b) and will continue to propagate there. The 'input' to the junction of the two tracks, coming from the left, is thus steered in to either the upper track or the lower track depending upon whether the current in the steering loop is either off or on.

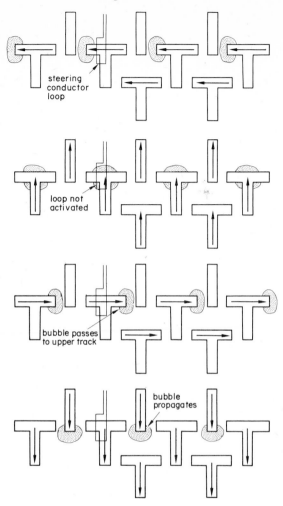

Figure 1.18(a). A branch in a bubble domain propagation track.

Figure 1.19 shows a bubble domain signal duplicator, or amplifier, described by Chang *et al.* (1972). The left hand side of this diagram shows one end of a closed loop propagation track in which a bubble domain signal circulates anticlockwise as the drive field, B_D, rotates clockwise. Bubbles enter the sub-system at the left, on the lower track, and leave by the upper track, also on the left. The object is to produce a duplicate signal of the contents of the closed track which could then be used as an output; for example, this duplicate signal could be led off to a destructive read-out system as we described in §1.4.4 above.

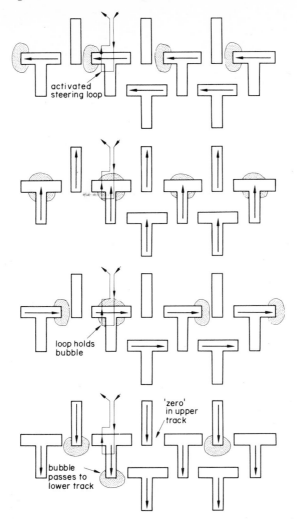

Figure 1.18(b). A steering conductor loop causes bubbles to be transferred to the lower track.

The bubble domain duplicator is of the greatest importance because it represents an equivalent of the power amplifier in familiar electronics. The duplicator has a power gain of two, because one signal enters the duplicator and two signals leave it, the original and the copy. The power input to the duplicator comes from the drive field.

An alternative approach to the problem of providing power gain in bubble domain devices is shown in figure 1.20. This sub-system uses a 'bubble domain power supply' in the form of a continuous stream of bubbles

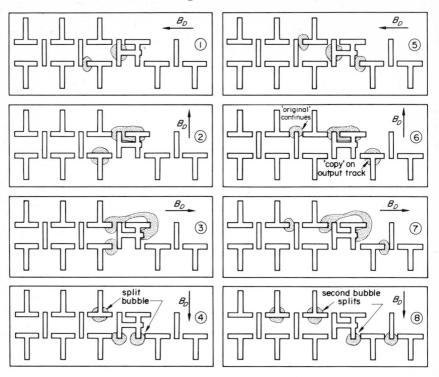

Figure 1.19. A bubble domain splitter. On the left hand side we see one end of a closed propagation track around which two bubbles are proceeding as the in-plane drive field rotates. Each bubble is split as we go from 3 to 4 and from 7 to 8, thus producing a copy on the output track of the signal stored in the closed propagation track.

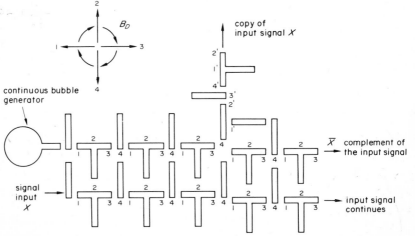

Figure 1.20. Interaction between the bubble domains propagating in two parallel tracks can determine what happens when a branch is made in one of these tracks.

entering the sub-system by the upper propagation track on the left of figure 1.20. After the third **T** element in this track, the propagation path is split and the bubbles would normally continue across the page. If, however, there is also a bubble domain signal in the lower track, propagation in the upper track will be switched into the path 1′2′3′4′ because of the mutual repulsion between a bubble in the lower track, making the third $4 \to 1$ transition, causing a bubble in the upper track to go from 4 to 1′ instead of also making the third $4 \to 1$ transition in the upper track.

The sub-system shown in figure 1.20 thus makes a copy of the input signal, preserves the input signal and also produces the complement of the input signal. Its power gain is, consequently, three and, from the point of view of a power amplifier, this sub-system would be particularly useful as a corner in a closed propagation track because it produces two output signals, X and $\overline{X}$, which could be used for destructive read–out of the data circulating in the closed track.

From the bubble domain logic point of view, the sub-system shown in figure 1.20 is an inverter; the output is $\overline{X}$ for an input signal X. If the continuous bubble generator is replaced with a second bubble domain signal, Y, as shown in figure 1.21, we have a logical AND gate. As shown in figure

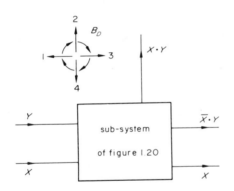

Figure 1.21. When the bubble domain generator of figure 1.20 is replaced by a binary bubble signal, Y, the outputs are as shown.

1.21, the output leaving the top of the sub-system is $X \cdot Y$ [(X) and (Y)] while that leaving the lower output is $\overline{X} \cdot Y$ ((not X) and (Y)). Note that there is an important difference between such a bubble domain logic gate and the familiar gates using semiconductors. This difference is in the propagation delay through the gate. In a semiconductor gate the propagation delay will be very small compared to the pulse width which would be used to represent a binary '1'. The bubble domain logic gate, in contrast, introduces a propagation delay of at least a few clock periods.

1.5 Conclusions

In this introduction we have seen what the magnetic bubble domain is and why there is such interest in the bubble domain device. At the time of writing, a considerable amount of highly successful work is going on in realising small data stores in bubble domains (Bosch *et al.* 1973) and in the feasibility of really large bubble memories of up to 10^8 bits (Almasi *et al.* 1971). These bubble domain devices are really complete subsystems because all the encoding and decoding, which is involved in the organisation of the memory, is done in the medium of bubble domains using transfer gates working in a very similar way to the simple illustration used here for figure 1.18.

The chapters which follow are intended to give the fundamental theory which is needed for the design and understanding of these bubble domain devices. We shall look at the bubble domain itself, statically to begin with, and then see what sort of magnetic materials are needed and how the material parameters determine the bubble size and effect stability. The dynamics of the bubble domain must be understood if devices are going to work at really high speeds and we shall see that the dynamic problems in bubble domain work can be quite fascinating in their complexity and have led to some very interesting theoretical and experimental work. All these dynamic studies eventually lead to the problems associated with the bubble domain propagation tracks and the very complex situation which results when we have to consider the bubble domains and their propagation tracks together.

References

Almasi, G. S., 1971, *IEEE Trans. on Magnetics*, MAG 7, 370.
Almasi, G. S., Keefe, C. E., and Terlep, K. D., 1972, High speed sensing of small bubble domains, *18th Conf. on Magnetism and Magnetic Materials*, Denver, Nov. 1972 (*AIP Conf. Proc.* No. 10, p. 207).
Almasi, G. S., Bouricius, W. G., and Carter, W. C., 1971, *AIP* Conf. Proc. No. 5, p. 225.
Bobeck, A. H., and Fischer, R. F., 1959, *J. appl. Phys.*, **30** (supplement), 43S.
Bobeck, A. H., 1967, *Bell Syst. tech. J.*, **46**, 1901.
Bobeck, A. H., Fischer, R. F., Perneski, A. J., Remeika, J. P., and van Uitert, L. G., 1969, *IEEE Trans. on Magnetics*, MAG 5, 544.
Bosch, L. J., Downing, R. A., and Rosier, L. L., 1973, *TEEE Trans. on Magnetics*, MAG 8, 481.
Bozorth, R. M., 1961, *Ferromagnetism*. Van Nostrand, Princeton, N.J., Chapter 12.
Broadbent, K. D., 1960, *IRE Trans. Elec. Comp.* EC9, 321.
Chang, H., Fox, J., Lu, D., and Rosier, L. L., 1972, *IEEE Trans. on Magnetics*, MAG 8, 214.
Galt, J. K., Matthias, B. T., and Remeika, J. P., 1950, *Phys. Rev*, **79**, 391.
Kittel, C., 1949, *Rev. mod. Phys.*, **21**, 541.
Nielson, J. W., and Dearborn, E. F., 1958, *J. Phys. Chem. Solids*, **5**, 202.
Perneski, A. J., 1969, *IEEE Trans. on Magnetics*, MAG 5, 554.
Remeika, J. P., 1956, *J. Amer. Chem. Soc.*, **78**, 4259.
Sherwood, R. C., Remeika, J. P., and Williams, H. J., 1959, *J. appl. Phys.*, **30**, 217.
Snoek, J. L., 1947, *New Developments in Ferromagnetic Materials*. Elsevier Publishing Co. Inc.
Strauss, W., and Smith, G. E., 1970, *J. appl. Phys.*, **41**, 1169.

2   Magnetostatics

In order to understand bubble domain devices, we need to know how to calculate the magnetic fields which are produced by magnetic domains. Provided that the magnetic domain pattern is defined, this is a problem in magnetostatics and, in this chapter, we shall look at some methods of solving the magnetostatic problem subject to the same assumption that was made in Chapter 1 when figures 1.4 to 1.9 were discussed. This assumption was that we are dealing with a magnetic layer having a very well-defined easy direction of magnetisation normal to its plane. The implications of this assumption will be considered later in this chapter and in Chapter 3 where magnetic crystals are considered.

The magnetostatic problems associated with bubble domains have been treated in great detail by a number of authors. Reference to this work will be made here, as completely as possible, but the intention of this chapter is to present these ideas from a rather different point of view in order to get a better picture of what is really happening in these domain problems.

2.1   The Magnetisation

2.1.1   *Representing the magnetisation by means of equivalent currents*

To understand the behaviour of magnetic materials, we must remember that their magnetisation is due to the magnetic moment of some, or all, of their constituent atoms being ordered. In a single crystal of magnetic material, this ordering simply means that all the atomic moments are aligned parallel to some particular crystallographic direction.

The field equations which we use in electromagnetism, Maxwell's equations, say nothing about the atomic nature of matter and this means that we can accurately represent the magnetisation of our material by means of a large number of elementary circulating currents. An example is shown in figure 2.1 where we consider a cylinder of magnetic material, height h and diameter D, which is uniformly magnetised in a direction parallel to its axis. The cylinder is to be imagined as being filled with a large number of small circulating currents, each occupying a small volume a^3, so that any cross-section of the cylinder will appear as shown in figure 2.2.

Because all the circulating currents, representing the atomic scale magnetic moments, are of the same intensity, i, figure 2.2 shows that adjoining mesh currents cancel one another out. They are equal and opposite. It is only at the surface of the cylinder, around its circumference, that we have any effective current and this leads to a most important result: the magnetisation of a uniformly magnetised body may be represented exactly by means of an equivalent surface current.

Magnetic Bubbles

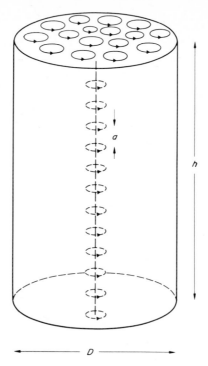

Figure 2.1. A uniformly magnetised cylinder. The atomic magnetic moments may be represented by small circulating currents, each occupying a volume a^3.

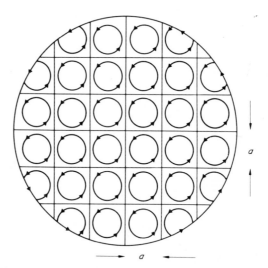

Figure 2.2. The elementary circulating currents cancel one another out everywhere except around the boundary of the magnet.

In other words, the uniformly magnetised cylinder produces a magnetic field, B, both inside and outside, which is identical to that produced by a solenoid of the same dimensions which is carrying a total surface current $i(h/a)$. The number, (h/a), is simply the total number of current loops stacked one above the other along the axis of the cylinder shown in figure 2.1. To make this a useful result, we only need to know the relationship between i and the magnetisation of the material, M. This is a matter of definition and the accepted one is that the magnetisation of a material is equal to its magnetic moment per unit volume. Again, by definition, the magnetic moment of a small current loop is equal to the circulating current multiplied by its area (Feynman 1964, pp. 14–28) and reference to figure 2.2 shows that the elementary magnetic moment which we are dealing with here has a value ia^2.

It follows that $M = (ia^2)/a^3$ is the relationship between the magnetisation and the elementary current representing it. The total surface current of our equivalent solenoid, given above as $I = i(h/a)$, is simply $I = Mh$.

2.1.2 *The magnetic field of a cylinder*

We can now evaluate the magnetic field of the uniformly magnetised cylinder, shown in figure 2.1, for two very simple cases. In the first case, consider a very long cylinder, $h \gg D$. This is equivalent to a very long solenoid and we know that the magnetic field inside such a solenoid is fairly uniform and is very nearly equal to

$$B = \mu_0 I/h \qquad (2.1)$$

where $\mu_0 = 4\pi \times 10^{-7}$, I is the total surface current, h is the length of the solenoid and the magnetic field, B, is directed along the axis. In the case of a long cylinder, magnetised uniformly to M along its length, we replace I by Mh, in equation (2.1), to obtain,

$$B = \mu_0 M, \qquad h \gg D \qquad (2.2)$$

as the field within the cylinder.

In contrast we shall look at a very short cylinder, $D \gg h$, for our second case. This is relevant to bubble domain devices because it represents the situation we shall find in a thin layer of material which is uniformly magnetised in a direction normal to its plane. Such a thin layer will produce a magnetic field which is identical to that of a thin coil, as shown in figure 2.3. The current flowing around the periphery is, again, $I = Mh$. The magnetic field at the centre of such a single turn, when $D \gg h$, is simply $\mu_0 I/D$, or in terms of M

$$B = \mu_0 M(h/D), \qquad D \gg h \qquad (2.3)$$

This is a very small field compared to that given by equation (2.2). The factor (h/D) is, in fact, the so-called demagnetising factor of a cylinder for which $D \gg h$ (Sommerfeld 1952, p. 82).

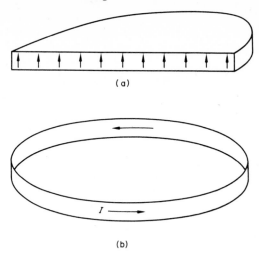

Figure 2.3. A thin circular layer of uniformly magnetised material (a) produces the same field as the single turn coil shown in (b).

A much larger magnetic field is found at the edge of the thin layer than at the centre. This can be found very simply by making use of the well-known Ampere's rule which tells us that the line integral around a contour of the vector B is equal to the current enclosed by the contour multiplied by μ_0, that is

$$\int_c B \cdot dl = \mu_0 I \qquad (2.4)$$

(Sommerfeld 1952, p. 13, Feynman 1964, p. 13–4).

At the extreme edge of the circular coil, shown in figure 2.3(b), the vector B is simply directed up one side of the current sheet and down the other side so that a sensible choice of contour for equation (2.4) follows the same path and will have a length $2h$. Along such a contour, B, will be parallel to dl and have very nearly constant magnitude so that the left hand side of equation (2.4) is simply $2hB$. As $I = Mh$ it follows that

$$B = \tfrac{1}{2}\mu_0 M \qquad (2.5)$$

at the edge of the thin layer, in contrast to the very small field at the centre given by equation (2.3).

In a typical bubble domain device we would be dealing with a thin layer for which h and D would be of the order of 5×10^{-6} m and 10^{-2} m, respectively. The value of M would be the saturation magnetisation of the material, M_s, and would be such that $\mu_0 M_s \approx 200 \times 10^{-4}$ T. It follows from equation (2.3) that the magnetic field at the centre of such a thin layer, due to its own magnetisation, would be only 0.1×10^{-4} T. Such a small field would be quite negligible in a practical device. The field is still very small as we approach the edge of the thin layer because a full analysis of this problem shows that B rises to the large value given by equation (2.5) within a distance of the order of h from the edge.

2.1.3 *A comment on the use of μ_0 and B*

Equation (2.4) is the form taken by Maxwell's integral equation for the particular case of the vacuum and its use here implies that we are assuming that a saturated magnetic material has a permeability μ_0. This is justified because the large relative permeability which is commonly associated with a magnetic material is only due to the motion of domain walls. The saturated material will have a permeability very close to μ_0. This assumption, and the representation of M by equivalent circulating currents, allows us to concentrate upon the magnetic field B, and the magnetisation, M, and so avoid the confusion which so often arises when the third magnetic vector H, is introduced. It is modern practice in magnetism to refer to B as the magnetic field and this fundamental point is considered in detail by Sommerfeld (1952, pp. 9–11, p. 218) and by Feynman (1964, ch. 13, ch. 14).

2.2 The Contraction of a Strip Domain in an Applied Bias Field

We are now in a position to consider the behaviour which was illustrated in figure 1.4 and shown in the micrographs, figures 1.5 and 1.6. In these figures, we saw that the effect of an externally applied magnetic field was to reduce the volume of material magnetised against the applied field while increasing the volume magnetised in the same direction as the applied field. We have to explain, quantitatively, how it is that the narrow strip domains, shown in figure 1.6, are held in stable equilibrium by means of the applied field and obtain a relationship between their width and the magnitude of the applied field.

2.2.1 *The force on the wall of a strip domain*

Figure 2.4 shows an isolated strip domain, width W, in a thin layer of thickness h. We shall assume that the thin layer is of considerable diameter so that we can neglect any effects due to its boundary by virtue of equation (2.3).

Figure 2.4 shows the equivalent circulating current model of the magnetisation in and around this strip domain and we can see that adjacent mesh currents cancel out, as in figure 2.2, except at the boundary or wall where they add up. This is exactly the same kind of effect which was discussed in connection with figure 2.2, concerning the periphery of the cylinder, except in the case of the strip domain the effective current at the wall is $2I$, that is $2Mh$, because the magnetisation, M, is reversed outside the strip domain instead of being simply zero as it was outside the cylinder.

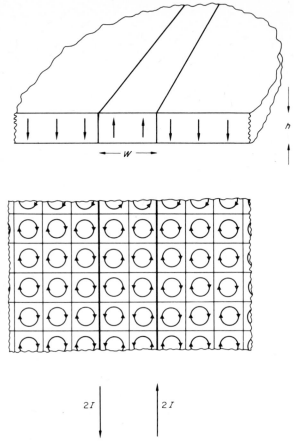

Figure 2.4. A strip domain is equivalent to two currents localised at the domain walls. The current is $2I$, compared to figure 2.3, because the magnetisation reverses at the boundary instead of just becoming zero.

It follows that the strip domain may be accurately represented by means of two parallel currents, as shown in figure 2.4. It is interesting to pause for a moment and ask what order of magnitude these equivalent currents will have in a typical bubble domain material. Suppose $h = 5 \times 10^{-6}$ m and $\mu_0 M \approx 200 \times 10^{-4}$ T. The equivalent current is then $\approx 0\cdot 2$ A. This is an interesting result when we remember that this current flows in a wall region $5\ \mu m$ high and, perhaps, only $0\cdot 1\ \mu m$ wide.

Having realised that the strip domain is identical to two parallel currents, its contraction in an externally applied field, B_0, follows at once. As shown in figure 2.5, each wall will experience a force per unit length $F = 2MhB_0$, because a current carrying conductor experiences a force in a magnetic field which is at right angles to both the current direction and the magnetic

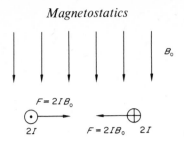

Figure 2.5. The force on the walls of the strip domain when a field B_0 is applied.

field direction and has a magnitude, per unit length, given by the product of the current and the field. To be exact

$$F = i \times B \qquad (2.6)$$

when the current, i, is uniformly distributed, so that the force is directed inwards for the situation shown in figure 2.5. This is the force which causes the strip domain to contract as the applied field is increased. The reason it contracts to some well-defined width can be seen from figure 2.6(a) where

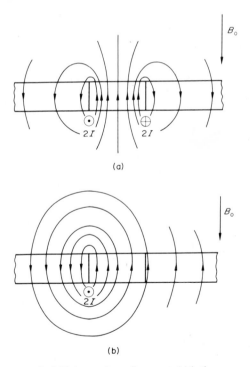

(a)

(b)

Figure 2.6. The magnetic field due to the wall currents (a) is the superposition of two fields, the left hand one is shown in (b). This field produces a repulsive force upon the right hand wall.

the magnetic field due to the two walls themselves is shown. This field is simply the superposition of the fields of the individual walls and one of these is shown in figure 2.6(b).

In figure 2.6(b) we can see that the field produced by the left hand wall at the right hand wall is in the opposite direction to B_0 and produces a force in the opposite direction to that shown in figure 2.5. The two walls repel one another. Furthermore, this repulsion increases as the distance between the walls decreases, whereas the force due to the external field is constant, so that it is obvious that the strip domain contracts until the two forces balance.

2.2.2 *The magnetic field produced by a strip domain*

In order to obtain a quantitative model for the contraction of the strip domain in an applied field, we have to calculate the magnitude of the mean magnetic field, $\bar{B}$, which the left hand wall shown in figure 2.6(b) produces at the right hand wall. We can do this very easily for two extreme cases by using Ampere's rule again. This was equation (2.4).

The first extreme case we consider is when $W \gg h$. This is illustrated in figure 2.7(a) which shows that, when $W \gg h$, the wall current, viewed from its neighbouring wall, looks more and more like a concentrated current, $2I$, the more W exceeds h. The field lines, at the right hand wall, appear more and more circular so that the length of the contour in equation (2.4) tends to $2\pi W$ and we have the result

$$\bar{B} = \mu_0 \frac{2I}{2\pi W} \qquad W \gg h \qquad (2.7)$$

for the field produced at the right hand wall by the left hand wall. Because $I = Mh$, it follows that

$$\frac{\bar{B}}{\mu_0 M} = \frac{1}{\pi(W/h)} \qquad W \gg h \qquad (2.8)$$

where we have normalised $\bar{B}$ to $\mu_0 M$ for convenience.

The second extreme case we shall consider is when $W \ll h$ and we are very close to the left hand wall. The situation is then that shown in figure 2.7(b) and we can see that the lines of constant field now lie very close to the left hand wall and, in the limit as $W \to 0$, they will have a length $2h$. From equation (2.4) again, it follows that

$$\bar{B} = \mu_0 \frac{2I}{2h} \qquad W \ll h \qquad (2.9)$$

and, as $I = Mh$,

$$\frac{\bar{B}}{\mu_0 M} = 1 \qquad W \ll h \qquad (2.10)$$

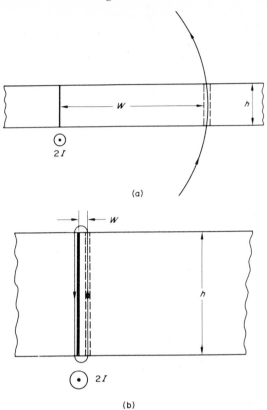

(b)

Figure 2.7. Calculating the field at the right-hand wall due to the effective current in the left-hand wall, for a wide strip domain (a) and a very narrow one (b).

Now equations (2.8) and (2.10) are exactly correct in their limits, $W \rightarrow 0$ and $W \rightarrow \infty$, so that we know that a plot of $\bar{B}/\mu_0 M$, as a function of W/h, will be of the form shown in figure 2.8; starting at $\bar{B}/\mu_0 M = 1$ and eventually following equation (2.8). The simplest *approximation* to this function is obviously

$$\frac{\bar{B}}{\mu_0 M} = \frac{1}{1 + \pi(W/h)} \tag{2.11}$$

which is, again, exactly correct at the two extremes and does, in fact, differ by only a few per cent from the exact solution,

$$\frac{\bar{B}}{\mu_0 M} = \frac{2}{\pi}\left[\tan^{-1}\left(\frac{h}{W}\right) - \frac{W}{2h}\log_e\left(1 + \frac{h^2}{W^2}\right)\right] \tag{2.12}$$

given by Kooy and Enz (1960) and by Bobeck (1967) in their consideration of the magnetostatics of the isolated strip domain. It is not easy, however,

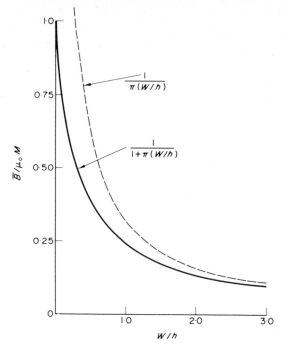

Figure 2.8. The wall field in a strip domain as a function of (W/h).

to see what kind of relationship equation (2.12) really is without either plotting it out or taking it to the extremes, $W \gg h$, $W \ll h$, whereupon it is identical to equations (2.8) and (2.10) and we would be led back to equation (2.11) again. In any case, equation (2.12) is obtained by assuming absolutely uniform magnetisation both inside and outside the domain and zero wall thickness. We shall see later that such assumptions may lead to errors at least as large as the difference between equation (2.12) and our approximation, equation (2.11), so that the use of a sensible approximation has everything to recommend it. Approximations of this kind were first proposed by Callen and Josephs (1971).

2.2.3   *Equilibrium of the strip domain in an applied field*

We can see that the strip domain is in stable equilibrium, as far as its width is concerned, from figure 2.9 where the mean wall field, $\bar{B}$, is compared with the applied bias field, B_0. Both fields have been normalised to $\mu_0 M$. For the particular value of B_0 chosen, the two fields cancel one another out at the point A, in figure 2.9, which corresponds to a value of $W/h \approx 0.8$. If the strip is narrower than this, $\bar{B}$ exceeds B_0 and the strip expands towards A. If the strip is too wide, B_0 predominates and the strip contracts towards A.

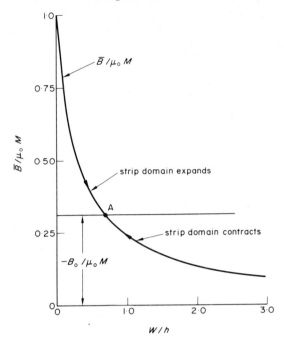

Figure 2.9. The strip domain is stable at the point A when the applied field is B_0 as shown.

It is interesting to note that, in this analysis, we have only considered the force on the strip domain wall due to the externally applied field, B_0, and the field, $\bar{B}$, due to the neighbouring wall. We have said nothing about the force upon the wall due to its own field. Figure 2.6(b) shows that, provided the wall is straight and infinitely long, its own field surrounds it symmetrically so that any forces due to this field balance one another and may be ignored. This is not a stable situation, however, because the effective wall current can be thought of as a current flowing in an almost perfectly elastic and mobile conductor. Such a current is subject to what is known as a 'kink instability' which, from a field point of view, may be seen to arise when the conductor is bent, as shown in figure 2.10. The magnetic field on the inside of such a bend increases slightly while that on the outside of the bend falls. Consequently, the magnetic forces no longer balance and the 'kink' increases in intensity. Such instabilities are well-known in plasmas and a concise and general treatment may be found in Landau and Lifshitz (1960, pp. 131–136). The kink instability is the explanation for the serpentine form of the strip domains shown in figure 1.6. The strip domains settle down to have a stable serpentine form in these materials because there is a small energy associated with the domain walls themselves, which we shall

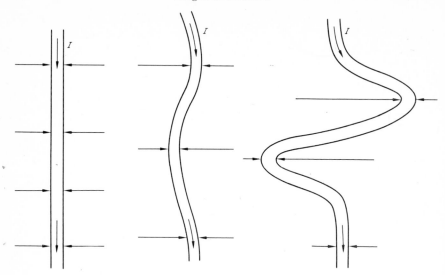

Figure 2.10. The kink instability of a linear conductor carrying a constant current. The magnitude of the force is indicated by the length of the arrows.

have to consider later in connection with the bubble domain. This additional energy term will be increased if the length of the wall increases so that it acts as a stabilizing force. The strip domains also lose their tendency to kink when the applied field B_0 is very high. This is because the two walls are brought very close together and the kink instability of one wall inhibits the instability of its neighbour, allowing the domain wall energy term to have a larger stabilizing effect.

2.3   The Magnetic Bubble Domain

We can now consider the magnetisation of the bubble domain itself. This was described briefly in § 3 of Chapter 1, where figure 1.10 presented the bubble domain as being held in equilibrium by the equal and opposite forces produced by the externally applied field and the internal field. An exact calculation of the internal field of the bubble domain can be made quite simply, the most elegant solution being that given by Bobeck (1967) who simplifies the problem by using a subtle co-ordinate transformation. This solution depends upon the same assumptions, which were mentioned at the end of § 2.2 above, that the domain walls have zero thickness and that the magnetisation is absolutely uniform.

We shall not repeat this exact calculation here because the result is not very useful, in the same way that equation (2.12) was not. Instead, we shall justify the use of a much simpler expression for the mean wall field of the bubble domain, as we did for the strip domain when we obtained equation

(2.11). This approach not only results in a much more useful mathematical model but also underlines some very fundamental differences between the strip domain and the bubble domain.

2.3.1 *The external force upon the bubble domain*

The magnetic bubble domain, as shown in figure 2.11, is a cylindrical domain, diameter D, passing right through a thin layer of magnetic material, thickness h. The magnetisation within the domain is uniformly directed upwards while the magnetisation in the surrounding material is directed downwards. We assume that the domain wall has negligible width.

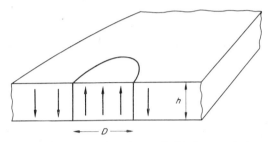

Figure 2.11. The magnetic bubble domain: a cylindrical magnetic domain, diameter D, which passes right through a magnetic layer of thickness h. The easy direction of magnetisation is very well defined in bubble domain materials and is normal to the plane of the layer.

Magnetostatically, the bubble domain may be modelled in the same way as the cylinder of figure 2.1 was, except that the equivalent wall current is now $2Mh$, as it was for the strip domain, because the magnetisation undergoes a complete reversal at the domain wall.

In an externally applied magnetic field, B_0, directed along the axis of the domain and opposed to the magnetisation inside the domain, there will be a total force upon the wall given by the product of the applied field, the wall current and the wall length. That is

$$F_0 = -(-B_0)(2Mh)(\pi D) \tag{2.13}$$

and this force will be radial and directed inwards so that the magnetic bubble will be compressed. Note that B_0 is a negative quantity as it is a field which is opposed to M. $-B_0$ is thus positive.

2.3.2 *The expanding force due to the field of the bubble domain*

The strip domain, considered in § 2.2, was found to have an expanding force due to its own field because the two walls repelled one another. We could consider the action of one wall upon the other very simply in that problem but the bubble domain problem has a rather fundamental difference. This is that there is only one wall, closed upon itself, and it is the action of the entire wall reacting upon any point on it which is now the equivalent of the, previously, very simple repulsion of two walls.

There is a very simple way of looking at this problem. Because the bubble domain is represented as a closed circuit, carrying the wall current $2Mh$, we can consider it as a circuit with inductance, L. The energy stored in an inductive circuit is

$$E = -\tfrac{1}{2}I^2 L \tag{2.14}$$

where the negative sign indicates that this is a stored energy. Equation (2.14) immediately tells us that, if the current is constant, there will be a force upon any circuit which will attempt to *increase the inductance* (Landau and Lifshitz 1960, pp. 131–136) and so reduce the energy E, given by equation (2.14). As the inductance of the circular current is a function of its radius, this means that there will be a radial force

$$F_r = -\frac{\partial E}{\partial r} = \frac{1}{2}I^2 \frac{\partial L}{\partial r} \tag{2.15}$$

which will cause the bubble domain to expand.

This force upon circuits, due to their liking for high inductance, is very familiar. It is this force which causes an iron core to slide into a solenoid which is carrying a constant current and is the same force responsible for the kink instability discussed in §2.2.3.

Let us now apply equation (2.15) to the bubble domain for two extreme cases. These will be the extremes $D \ll h$ and $D \gg h$ because, in both cases, we can say something about the inductance.

In the case of a long thin bubble domain, $D \ll h$, we know that the internal B field is like that of a long solenoid and is given by $\mu_0 I/h$. The magnetic flux through the solenoid is then

$$\phi = (\mu_0 I/h)(\pi D^2/4) \tag{2.16}$$

so that the inductance is

$$L = \frac{\phi}{I} = \pi\mu_0 r^2/h \tag{2.17}$$

where we replace D in equation (2.16) by $2r$ to allow us to use equations (2.15) and (2.17) to obtain the force

$$F_r = (\tfrac{1}{2}I^2)(2\pi\mu_0 r/h) \tag{2.18}$$

We now substitute the wall current, $I = 2Mh$, and write equation (2.18) in the same form as equation (2.13)

$$F_r = (\mu_0 M)(2Mh)(\pi D) \tag{2.19}$$

so that, comparing equations (2.13) and (2.19) shows that F_r and F_0 will be in equilibrium when the applied field, B_0, is equal to $\mu_0 M$ and opposed to it. In the same terminology used for the strip domain, in § 2.2, we can say that we have a mean effective field at the wall of the bubble domain, due to its own magnetisation, given by

$$\frac{\bar{B}}{\mu_0 M} = 1 \qquad D \ll h \tag{2.20}$$

in analogy to equation (2.10).

We now consider the other extreme of a bubble having a diameter very much greater than its length, $D \gg h$. This can be thought of as a single turn coil of a large diameter and we know, from our consideration of this in § 1.2, that the magnetic field is concentrated around its edge and we might expect the inductance to increase linearly with the circumference. This is very nearly the case, but an exact analysis, given for example by Sommerfeld (1952, p. 112), shows that

$$L = \mu_0 r \log_e (4\pi r/h) \tag{2.21}$$

or that the inductance does not increase simply with the radius but that

$$\frac{\mathrm{d}L}{\mathrm{d}r} = \mu_0 [1 + \log_e (4\pi r/h)] \tag{2.22}$$

showing a rate of increase which increases logarithmically, that is very slowly, with the radius itself.

This simple result, equation (2.22), is very fortunate because it means that we can represent $\mathrm{d}L/\mathrm{d}r$ by means of a constant over quite a wide range of D/h with very little error. That is, we shall let

$$\frac{\mathrm{d}L}{\mathrm{d}r} = \mu_0 K \tag{2.23}$$

where K is the 'constant', $[1 + \log_e (4\pi r/h)]$, to be chosen for the best approximation in our final result. Equation (2.15) then becomes

$$F_r = \tfrac{1}{2}(I^2)(\mu_0 K) \tag{2.24}$$

and this will be written in the same form as equation (2.13), after substituting the effective wall current, $I = 2Mh$, to obtain

$$F_r = \frac{\mu_0 M}{\pi/K(D/h)}(2Mh)(\pi D) \tag{2.25}$$

Comparing equations (2.25) and (2.13) shows that there is a magnetic field at the wall of the bubble domain,

$$\frac{\bar{B}}{\mu_0 M} = \frac{1}{\pi/K(D/h)} \qquad D \gg h \tag{2.26}$$

due to its own magnetisation, represented here by its effective wall current, which causes the expanding force given by equation (2.25). Equation (2.26) is, then, analogous to our equation (2.8) for the strip domain and applies when $D \gg h$ and for a suitable choice of K.

We may now combine equations (2.20) and (2.26), in the same way as we combined equations (2.10) and (2.8) to form equation (2.11). This time, the combination will yield,

$$\frac{\bar{B}}{\mu_0 M} = \frac{1}{1 + \pi/K(D/h)} \tag{2.27}$$

which becomes identical to equation (2.20) when $D \ll h$ and identical to equation (2.26) when $D \gg h$. The only remaining problem is to choose a value for K, which from equations (2.22) and (2.23) should be given by

$$K = 1 + \log_e 4\pi r/h \tag{2.28}$$

where the value of r/h to be used in equation (2.28) should be the value where we would wish our approximation to be valid. This would be where $D/h \approx 5$, which would make $\pi/K \approx 0.7$, because we shall see that we are only concerned with bubble domains where $0 < D/h < 3$. Callen and Josephs (1971) chose a value of $\pi/K = 0.75$ and showed that this gave a good approximation to the mean wall field of the bubble domain over the range $0 < D/h < 10$. For a more restricted range of D/h, however, a better choice for the constant in equation (2.27) is $\pi/K = 0.726$. This gives an excellent approximation to the exact solution and a comparison between the two is shown in figure 2.12. It can be seen that the approximation fits the exact solution, given by Bobeck (1967), at values of $D/h = 0$, 1.0 and 3.0. It is interesting to compare Bobeck's exact solution, which is

$$\frac{\bar{B}}{\mu_0 M} = \frac{2}{\pi}[(1 + D^2/h^2)^{1/2}E(k) - D/h] \tag{2.29}$$

where $E(k)$ is the complete elliptic integral of the second kind and the argument, k, is

$$k = (D/h)/(1 + D^2/h^2)^{1/2} \tag{2.30}$$

with our very simple approximation, equation (2.27), which is reduced to

$$\frac{\bar{B}}{\mu_0 M} = \frac{1}{1 + 0.726(D/h)} \tag{2.31}$$

when the best value of (π/K) is substituted.

It might be thought that we had now solved the problem of a bubble domain in equilibrium with an applied field and that we could simply equate $\bar{B}$ to the applied field B_0, as we did for the strip domain in figure 2.9. For example, if we considered the situation shown originally in figure 1.9, where we had 6 μm bubble domains in a 10 μm thick garnet film, and consulted figure 2.12, it might appear that these bubble domains, for which $D/h = 0.6$, would be held in equilibrium by an applied field, $B_0/\mu_0 M \approx 0.7$. This, however, would not be true. In the first place, the bubble domains shown in

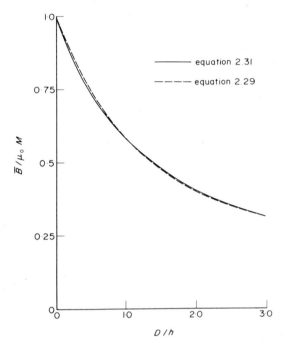

Figure 2.12. A comparison between Bobeck's exact solution for the mean field at the wall of a bubble domain, equation (2.29), and the approximate solution, equation (2.31).

figure 1.9 are not isolated and the field of neighbouring bubbles must be taken into account. This problem will be considered in Chapter 6, where we shall look at bubble domain arrays. In the second place, there is a further force upon the bubble domain which does not appear in the case of the strip domain, and this is due to the energy associated with the domain wall. This problem is considered in the next section.

2.3.3 *The contracting force due to the domain wall energy of a bubble domain*

A fundamental difference between the strip domain and the bubble domain was indicated at the beginning of § 2.3.2. This was that the strip domain could be considered very simply as two domain walls which repelled one another while the bubble domain involved only one closed wall and we had to consider the reaction of this wall upon itself.

Another fundamental difference between the two kinds of domain concerns the length of the domain walls. In the infinite strip domain the length of the two walls was constant. In fact, we considered the problem on a per unit length basis. The length of the domain wall around the bubble domain, however, is changing as its diameter changes and this introduces an entirely new term into our equations because this wall has an energy density in its own right and, as the length of the wall changes, so will the total energy of the bubble domain.

We shall not look closely at the origin of this domain wall energy until Chapter 3, where the magnetic crystals which concern us will be considered. It is sufficient at this point to realise that there is an energy density associated with this wall, and this may be seen by considering the simplest possible wall structure, which is shown in figure 2.13, where the wall is a simple

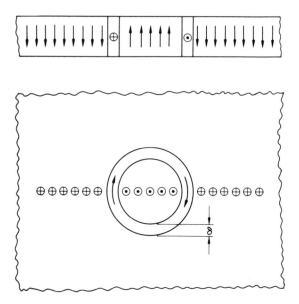

Figure 2.13. The simple Bloch wall which may surround a bubble domain.

Bloch wall and the magnetisation twists smoothly around the radial vector from being upwards within the domain to downwards outside. Within the wall itself, the magnetisation must be directed away from the easy direction, which is normal to the plane of the film in a bubble material, so that it is in a higher energy state. It follows that there must be a positive energy density associated with the wall.

The very simple wall structure shown in figure 2.13, in which the magnetisation circulates around the wall, is probably never found in nature because it would imply that there are two possible kinds of bubble domain; left-handed and right-handed, depending upon the relative sense of *M* within the bubble itself and within its wall. A symmetrical structure, in which the magnetisation within the wall alternated its direction of circulation would be more likely and is shown in figure 2.14. Each segment of the wall would be terminated by a small region in which the magnetisation would probably turn along the radius vector (Malozemoff 1972).

In either case, figure 2.13 or figure 2.14, there would be a positive energy density associated with the volume of the wall which could be treated as

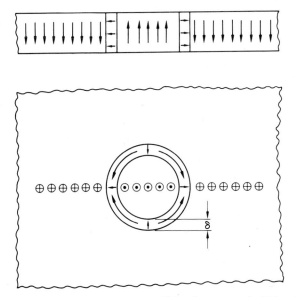

Figure 2.14. A more likely wall structure would involve a reversal of *M*, periodically around the circumference.

approximately constant. The relationship between this constant energy density and the physical parameters of the crystal will be considered in Chapter 3. At this stage, we shall represent this energy simply by a constant, σ_w Joules/m^2, which will be the mean energy density of the wall region multiplied by its thickness. From this standpoint, the total energy of the wall of a circular bubble domain of the kind shown in figures 2.13 and 2.14 will be

$$E_w = \sigma_w \pi D h \tag{2.32}$$

and will cause a radial contracting force upon the bubble domain given by

$$F_w = -\frac{\partial E_w}{\partial r} = -2\pi h \sigma_w \tag{2.33}$$

this being obtained from equation (2.32) by the same principle which yielded equation (2.15) from equation (2.14).

It is convenient to represent this wall force, F_w, by means of an effective field, B_w, because we shall have to consider it in comparison to the force due to the bias field, equation (2.13) and that due to the mean field at the wall, equation (2.31). This may be done very easily by asking what field, B_w, will produce a total force F_w upon a wall of length πD carrying our effective wall current, $2Mh$. The answer is

$$F_w = (B_w)(2Mh)(\pi D) \tag{2.34}$$

so that, from equation (2.33), B_w is given by

$$\frac{B_w}{\mu_0 M} = \frac{-\sigma_w}{\mu_0 M^2 h}(D/h)^{-1} \tag{2.35}$$

a negative field, as was the bias field, B_0, which tends to contract the domain and becomes more and more significant as the bubble diameter decreases and $(D/h)^{-1}$ becomes very large.

2.3.4 *The equilibrium of the magnetic bubble domain*

For the simple case in which the wall energy, σ_w, may be treated as a constant, we can now write down the equilibrium equation of the bubble domain as

$$\frac{B_0}{\mu_0 M} + \frac{B_w}{\mu_0 M} + \frac{\bar{B}}{\mu_0 M} = 0 \tag{2.36}$$

the contracting fields, B_0 and B_w, being negative, or opposed to the direction of M within the bubble, while the wall field $\bar{B}$ is positive. Substituting equations (2.31) and (2.35) into equation (2.36) then yields the condition

$$\frac{1}{1 + 0.726(D/h)} - \frac{(\sigma_w/\mu_0 M^2 h)}{(D/h)} = \frac{-B_0}{\mu_0 M} \tag{2.37}$$

which relates the bubble diameter to the applied field.

It is usual practice to simplify the following calculations by introducing a 'characteristic length' for the magnetic crystal under consideration which is defined as

$$\lambda = \sigma_w/\mu_0 M^2 \tag{2.38}$$

and has a value of the order of $1\,\mu\text{m}$ in most practical bubble domain materials.

If we now plot the left hand side of equation (2.37), a set of curves is obtained for various values of (λ/h) as shown in figure 2.15 and it is clear that there are two possible solutions to equation (2.37) where the curves take a

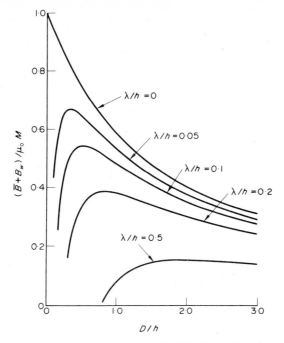

Figure 2.15. Showing the sum of the mean wall field of the bubble domain and the effective field due to the wall energy, for various values of λ/h.

value $(-B_0/\mu_0 M)$. A particular case is shown in figure 2.16 where we have chosen $\lambda/h = 0.1$ and $(-B_0/\mu_0 M) = 0.5$. The solutions are points 1 and 2. It is clear that point 2 should be the stable solution for the same reasons that point A in figure 2.9 was stable. Point 1, in figure 2.16, is unstable, in contrast, because if the bubble diameter departs from point 1 upon either side, a force results which tends to increase the deviation instead of restoring it.

2.3.5 The bubble domain collapse and run-out fields

If we consider figure 2.16 again, and imagine that the applied field, $-B_0/\mu_0 M$, is increasing gradually from the value 0.5 shown, it is clear that the two solutions, 1 and 2, will approach one another and coalesce when we reach the maximum of the curve shown. It is at this point that the bubble domain will collapse or burst and we can find the value of the collapse field, B_{COL}, and collapse diameter, D_{COL}, very simply by differentiating equation (2.37), with respect to (D/h) and equating the result to zero.

Let us write equation (2.37) as

$$\frac{1}{1 + q(D/h)} - \frac{(\lambda/h)}{(D/h)} = \frac{-B_0}{\mu_0 M} \qquad (2.39)$$

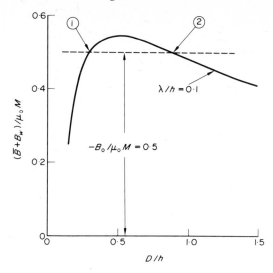

Figure 2.16. Equilibrium of the bubble domain with an applied field B_0. Only solution 2 is stable.

where we have replaced 0·726 by q for conciseness, and substituted equation (2.38). Differentiating equation (2.39) and equating to zero then yields

$$\frac{-q}{[1 + q(D_{COL}/h)]^2} + \frac{(\lambda/h)}{(D_{COL}/h)^2} = 0 \qquad (2.40)$$

which relates (D_{COL}/h), the collapse diameter, to (λ/h) which defines the material and layer thickness we are considering. Equation (2.40) is a simple quadratic for (D_{COL}/h) and the positive root

$$(D_{COL}/h) = \frac{(\lambda/h)^{1/2}/q^{1/2}}{[1 - q^{1/2}(\lambda/h)^{1/2}]} \qquad (2.41)$$

gives the value we require.

Substitution of equation (2.41) into equation (2.39) then gives the value of the collapse field in terms of (λ/h) as

$$(-B_{COL}/\mu_0 M) = [1 - q^{1/2}(\lambda/h)^{1/2}]^2 \qquad (2.42)$$

where q = 0·726.

Before evaluating equations (2.41) and (2.42) we must consider a further limit on the diameter of the isolated bubble domain when it is held in equilibrium by an applied field. This is the diameter at which it will 'run out' or lose its well defined circular shape and run out into a strip domain. The solution to this problem, in conjunction with the problem solved above, then gives us the range of possible diameters which can be maintained in some proposed bubble domain device which uses a particular material and layer thickness.

A glance back at figure 1.8, where the bubble domain was first introduced, will show that bubble domains can co-exist with short strip domains in a given bias field. A small increase in bias field will cause these short strip domains to contract into bubble domains and, similarly, a small drop in the bias field will cause the bubbles to run out into strips. Figure 1.8 shows a mixture of the two situations mainly because every domain shown there is influenced by the field of the surrounding domains. In the case of one isolated domain, the transition from strip to bubble, and vice versa, is well defined. The co-existence of bubbles and strips is illustrated in figure 2.17

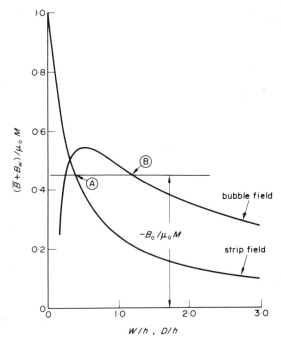

Figure 2.17. The curve originally shown in figure 2.9 and the curve from figure 2.15, for $\lambda/h = 0.1$, superimposed. It can be seen that strip domains, A, and bubble domains, B, may coexist in the same applied field B_0.

where the curve shown originally in figure 2.9, for the wall field of the strip domain, and one of the curves, for $\lambda/h = 0.1$, from figure 2.15, have been superimposed. It can be seen that infinitely long strips, point A, and bubbles, point B, can exist supported by the same field. The strip width in this case will be about one third of the bubble diameter, very much as it appears to be in figure 1.8.

The problem is further illustrated by figure 2.18 which shows the range of diameters which an isolated bubble may have between collapse and run out and its degeneration after run-out into an infinite strip domain. We have already obtained expressions for the field and diameter at which bubble collapse will occur and our problem now is to calculate the run-out field, B_{RO}, and the run-out diameter, D_{RO}.

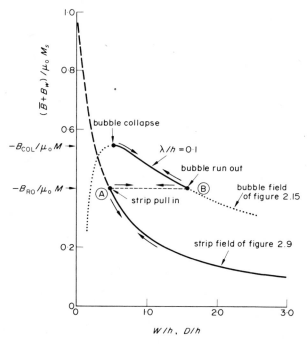

Figure 2.18. Defining the points of strip pull-in, bubble run-out and bubble collapse.

To do this, we again model the domain by means of its equivalent wall current, $2Mh$. In figure 2.19 we show the dumbbell shaped strip terminating in two bubble domains. We separate these in (c) and assume that they are in equilibrium with the applied field so that the forces on the short strip domain must be the same as those upon the imaginary rectangular domain shown in (d) as a rectangular current loop, width W and length l. This simple superposition is possible because the problem is linear.

We now calculate the critical field at which this rectangular strip will contract and propose that this is the same field at which the bubble domain will run-out. In other words, we are considering the situation which lies between the points A and B in figure 2.18.

There are two forces which will tend to make the rectangular domain contract in length. Its width may be assumed to be approximately constant.

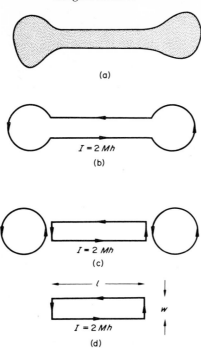

Figure 2.19. The short strip domain, see figure 1.8, is sketched in (a) and may be idealised as a current loop, (b), which is the superposition of two bubbles and a rectangular domain. To understand the stability of the short strip we need only consider the rectangular current loop, (d).

Considering one end of the rectangle shown in figure 2.19(d), the first of these contracting forces is due to the tension in the two long walls. These have an energy

$$E_w = \sigma_w h l \tag{2.43}$$

and, because there are two such walls, the total contracting force will be

$$F_w = -2 \frac{\partial E_w}{\partial l} = -2\sigma_w h \tag{2.44}$$

The second contracting force is due to the action of the applied field B_0 upon the end of the domain. This is treated as a conductor carrying the effective wall current $2Mh$, repeating the argument of §2.1, so that the contracting force due to this will be

$$F_{B_0} = (B_0)(2Mh)(W) \tag{2.45}$$

where we should note again that B_0 is a negative quantity.

The force which tends to make the rectangular domain expand is, again, its attempt to increase its inductance; the same effect which was discussed at length in connection with equations (2.15). To calculate this, we need to know the inductance per unit length of a pair of parallel current carrying strips, height h and separation W. This can be obtained at once from our approximate expression for the magnetic field of such a strip, which was given as equation (2.11). Integrating this expression, over the width of the strip, for both sides, leads to the inductance per unit length being given by

$$\frac{\mathrm{d}L}{\mathrm{d}l} = \frac{\mu_0}{\pi} \log_e \left[1 + \pi(W/h)\right] \tag{2.46}$$

so that, following the argument of equation (2.15), the expanding force on the rectangular domain is

$$F_e = \tfrac{1}{2}(2Mh)^2 \frac{\mu_0}{\pi} \log_e \left[1 + \pi(W/h)\right] \tag{2.47}$$

The answer to our problem of when the strip domain will pull-in, and, similarly, when the bubble domain will run-out, is thus given by setting the sum of all the forces on the rectangular domain to zero, that is

$$F_w + F_{B_0} + F_e = 0 \tag{2.48}$$

and solving for B_0, the applied field, which will then have the value B_{RO}, the run-out field with which we are concerned. To do this, we must first eliminate W/h from equations (2.45) and (2.47). This can be done very easily because we know that the strip domain is in equilibrium with the run-out field, at the point A in figure 2.18, so that, from equation (2.11)

$$\frac{-B_{RO}}{\mu_0 M} = \frac{1}{[1 + \pi(W/h)]} \tag{2.49}$$

may be substituted into equations (2.45) and (2.47) and these combined with equation (2.44) to give equation (2.48) as

$$-2\sigma_w h - \frac{2}{\pi}\mu_0 M^2 h^2 \left[1 - \left(\frac{-B_{RO}}{\mu_0 M}\right)\right] + \frac{2}{\pi}\mu_0 M^2 h^2 \log_e \left(\frac{\mu_0 M}{-B_{RO}}\right) = 0 \tag{2.50}$$

Equation (2.38) is then used to introduce the characteristic length, λ, into this equation and the final result is obtained as

$$\lambda/h = \frac{1}{\pi}\left\{\log_e \left(\frac{\mu_0 M}{-B_{RO}}\right) - \left[1 - \left(\frac{-B_{RO}}{\mu_0 M}\right)\right]\right\} \tag{2.51}$$

which is solved numerically to give values of the run-out field for given values of λ/h.

We now have our two essential results for the theory of bubble domains—equations (2.42) and (2.51), which give the collapse and run-out fields for a given material; that is, for a given value of λ/h. These results are plotted in figure 2.20. Despite the approximations we have made in this analysis, the

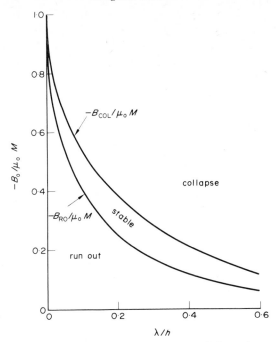

Figure 2.20. The bubble domain collapse and run-out fields for various values of λ/h.

results shown in figure 2.20 agree, to within a few percent, with those given by Thiele (1969, 1970) who analysed the problem with remarkable rigour.

Figure 2.20 shows the range of bias fields over which stable bubble domains may exist in a layer with a given value of λ/h. It can be seen that the ratio of B_{COL} to B_{RO} increases as λ/h increases and that, for a given value of $\mu_0 M$, the magnitude of the bias field becomes smaller as we increase λ/h. From this, it might appear sensible to use very thin layers for magnetic bubble domain devices because the absolute value of the bias fields required would be reduced and so would the tolerance on this bias field. We find, however, that this may not be a good argument when we look at the values of bubble diameter which are involved.

2.3.6 *The collapse and run-out diameters and the optimum value of λ/h*

The run-out and collapse diameters are easily found by substituting our results for the run-out and collapse fields back into the equilibrium equation (2.39). This has already been done for the collapse diameter and the result given as equation (2.41). In the same way, we can substitute the solution, $-B_{RO}/\mu_0 M$, of equation (2.51) back into equation (2.39), given the value of λ/h, and solve for D_{RO}/h. This can only be done within the computation which solves equation (2.51), however, because $-B_{RO}/\mu_0 M$ cannot be written explicitly in terms of λ/h.

Figure 2.21 shows the results of these calculations and we see at once that the range of stable diameters does increase very rapidly with λ/h but that the values of D/h shown are mainly well above unity. This would not normally be the requirement in a bubble domain device because the external field of a bubble domain with $D/h \gg 1$ will be very small and this will make

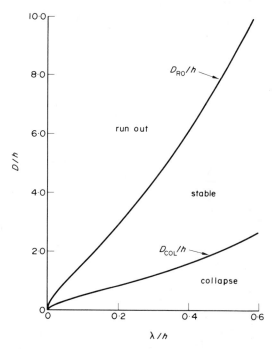

Figure 2.21. The bubble domain collapse and run-out diameters normalised to layer thickness, h.

detection difficult. On the other hand, bubbles with $D/h \ll 1$ would be unsuitable because these would not interact very well with the magnetic overlay patterns described in Chapter 1 which are normally put down upon only one side of the magnetic layer so that the interaction between an overlay and a tall thin bubble domain would be asymmetric in that the top of the bubble would experience a much stronger overlay interaction than the bottom.

A better understanding of the situation may be seen from figure 2.22 where we have presented the results again but, this time, the run-out and collapse diameters have been normalised to the characteristic length, λ. The value of λ is a constant for a given material at a given temperature and figure 2.22 shows that the stable range of bubble diameters, between D_{RO} and D_{COL}, is centred around 9λ to 10λ over a wide range of λ/h. The very

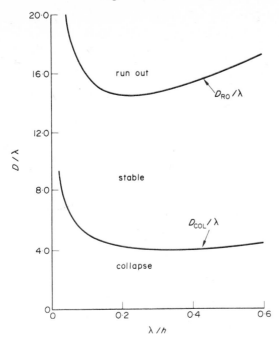

Figure 2.22. When the collapse and run-out diameters are normalised to λ, the characteristic length, which is constant for a given material, it can be seen that the mean bubble diameter is around 9λ for a wide range of λ/h. Very small values of λ/h allow larger bubbles.

weak minimum in D_{COL}/λ which occurs near $\lambda/h = 0.35$ would not justify choosing this value of λ/h as an optimum value, which would give a minimum in the absolute bubble diameter and thus a maximum in the bit density. Comparison of figures 2.21 and 2.22 would suggest that, given a material of a known λ, the thickness, h, would be adjusted so that λ/h lay between 0.1 and 0.2 to make the absolute bubble diameter about 10λ and the bubble shape such that D/h was around 1.0 to 1.5.

In practical bubble domain devices, we usually find just this kind of choice for λ/h, pushed, in fact, towards even lower values of λ/h than we might expect simply because the actual value of λ in many of the garnet materials is too small and a bubble diameter around 10λ would be difficult to deal with. Almasi *et al.* (1972), for example, describing their device work with particular reference to detectors, used 4 μm diameter bubbles in 4.7 μm thick layers of $Eu_{0.6}Y_{2.4}Fe_{3.9}Ga_{1.1}O_{12}$. The experimental devices described by Chang *et al.* (1972) were fabricated on 13 μm thick layers of $Gd_{2.24}Sm_{0.1}Tb_{0.66}Fe_5O_{12}$ which had a value of $\lambda = 0.33$ μm so that the value of λ/h was only 0.025. Such a small value of λ/h, as figure 2.22 shows, allowed stable bubbles with a diameter of about 16λ to be used and this was large enough to suit the overlay patterns being used in these devices. An early shift register described by Danylchuk (1971) used bubbles with $D/h = 0.5$ in $Gd_{2.34}Tb_{0.66}Fe_5O_{12}$, which suggests a value of λ/h below 0.08. The sub-micron bubble domain shift registers described by Plaskett *et al.* (1972) using $Eu_2YFe_5O_{12}$ all involved films for which λ/h was below 0.1. It is clear from these reports that it has not usually been possible to work with the smallest bubbles which a given material may be able to support.

2.4   Conclusions

In this chapter we have studied the magnetostatics of isolated strip and bubble domains and the way in which these domains may be held in static equilibrium with an applied bias field which is both uniform in space and in a direction which is antiparallel to the direction of the magnetisation within the domain. These calculations have led to results which tell us the limiting values of bubble domain diameter, figures 2.21 and 2.22, and bias field, figure 2.20, which would be found under such ideal conditions. In the more complicated situation found in a bubble domain device, where the bias field, B_0, is deliberately distorted by means of a magnetic overlay pattern and where an in-plane field is also applied, the limiting values of applied field and bubble diameter will depend upon a number of other factors which have not been considered so far. The situation is again complicated when we remember that the bubble domain device is dynamic and the bubble domain within the device must be stable under dynamic conditions as well as static conditions.

Nevertheless, the ideal static situation is very important as a first step in understanding device limitations and as a means of evaluating materials. A considerable literature has grown up, over the past few years, on the magnetostatic problems connected with bubble domains and more references will be given in Chapter 3 where some of the measurement techniques which have been worked out to evaluate materials will be described. These techniques often involve bubble domain arrays or periodic strip patterns, in contrast to the isolated domains which we have considered so far.

The literature on magnetic domain theory is extensive and perhaps the most comprehensive and critical review may be found in the books by Brown (1962, 1963). An early paper by Kooy and Enz (1960) considered strip domains and cylindrical domains theoretically and experimentally in $BaFe_{12}O_{19}$, which was shown to support sub-micron bubble domains, although the very high value of the saturation magnetisation in this material makes it unsuitable for any device applications. As we described in Chapter 1, it was Bobeck (1967) who first suggested the device potential of magnetic materials which are so strongly anisotropic that layers may be grown or cut so that the easy direction of magnetisation is normal to the surface of the layer and he gave a very interesting treatment of the magnetostatic problems involved. The stability of bubble domains was not considered in these earlier papers, however, but was covered in detail by Thiele (1969, 1970). Callen and Josephs (1971) introduced simple approximations into Thiele's theory to obtain the explicit expressions for bubble collapse diameter and collapse field which are given here as equations (2.41) and (2.42) but they did not consider the run-out problem in the light of these approximations as their work was directed towards dynamic problems. Cape and Lehman (1971) also gave a treatment of isolated strip and bubble domains in their paper which was mainly concerned with parallel strip patterns and bubble domain arrays.

Finally, we must say something about the effect of our idealisations and assumptions. These have been given at the beginning of this chapter—that the layer of magnetic material has a very well defined easy direction, normal to its plane, and that the width of the domain wall is negligible. We shall see in the next chapter that both these conditions are satisfied when the uniaxial anisotropy energy of the material is very high. In order to have an easy direction of magnetisation normal to the surface of a thin layer it is necessary to have K_u, the uniaxial anisotropy energy, greater than the magnetostatic energy, $(\mu_0 M)M/2$, where M is the saturation magnetisation of the film and it has become common practice to quote a 'quality factor', Q, for bubble domain materials given by

$$Q = \frac{2\mu_0 K_u}{(\mu_0 M)^2} \tag{2.52}$$

The value of Q in most bubble domain materials is about 5 and we shall see in Chapter 3 that the domain wall widths in most materials are the order of 0·1 μm. It follows that we can neglect the wall width, provided the bubble domain is a few microns in diameter, and also assume that the domain walls are straight and parallel to the easy direction of magnetisation, provided $D/h \approx 1$ or $D/h > 1$, because of the high anisotropy. In connection with these assumptions it should be noted that a wall width of 0·1 μm will introduce an error into our calculation of $\bar{B}/\mu_0 M$, equation (2.31), which will be greater than the difference between our approximation, equation (2.31), and the exact solution given as equation (2.29).

References

Almasi, G. S., Keefe, G. E., and Terlep, K. D., 1972, High speed sensing of small bubble domains, *18th Conf. on Magnetism and Magnetic Materials*, Denver, Nov. 1972 (*AIP Conf. Proc.* No. 10, p. 207).

Bobeck, A. H., 1967, *Bell Syst. Tech. J.*, **46**, 1901.

Brown, W. F. Jr., 1962, *Magnetostatic Principles in Ferromagnetism*. North Holland Pub. Co., Amsterdam.

Brown, W. F. Jr., 1963, *Micromagnetics*. Interscience Publishers, New York.

Callen, H., and Josephs, R. M., 1971, *J. Appl. Phys.*, **42**, 1977.

Cape, J. A., and Lehman, G. W., 1971, *J. Appl. Phys.*, **42**, 5732.

Chang, H., Fox, J., Lu, D., and Rosier, L. L., 1972, *IEEE Transactions on Magnetics*, MAG 8, 214.

Danylchuk, I., 1971, *J. Appl. Phys.*, **42**, 1358.

Feynman, R. P., 1964, *Lectures on Physics*, vol. II. Addison-Wesley Publishing Co. Inc., Reading, Mass.

Kooy, C., and Enz, U., 1960, *Philips Res. Repts.*, **15**, 7.

Landau, L. D., and Lifshitz, E. M., 1960, *Electrodynamics of Continuous Media*. Pergamon Press, Oxford.

Malozemoff, A. P., 1972, *Appl. Phys. Letts.*, **21**, 149.

Plaskett, T. S., Klokholm, E., Hu, H. L., and O'Kane, D. F., 1972, Magnetic Domains in $(EuY)Fe_5O_{12}$ Films on $Sm_3Ga_5O_{12}$ Substrates, *18th Conf. on Magnetism and Magnetic Materials*, Denver, Nov. 1972 (*AIP Conf. Proc.* No. 10, p. 319).

Sommerfeld, A., 1952, *Electrodynamics*. Academic Press, New York.

Thiele, A. A., 1969, *Bell Syst. Tech. J.*, **48**, 3287.

Thiele, A. A., 1970, *J. Appl. Phys.*, **41**, 1139.

3 Magnetic Bubble Domain Materials and Their Characteristics

In this chapter, we shall consider the physics of bubble domain materials, the preparation of these materials and, finally, their characterisation or the measurement of their properties. Only static properties and measurements will be considered in any detail at this stage because the dynamic properties and the measurement of dynamic effects will be considered in Chapter 4. The preparation, or crystal growth, problems will only be outlined here because this field has a vast and rapidly growing literature to which reference must be made for the latest technical developments. Our purpose here is to consider what kind of magnetic materials might be suitable for bubble domain work and we shall see that the theoretical work given in Chapter 2 will put quite strict constraints on our choice of these materials.

3.1 The Physics of Bubble Domain Materials

The physics of bubble domain materials is the physics of magnetic materials, restricted to the special case of thin layers which have a well defined easy direction of magnetisation normal to the plane of the layer. In order to see how the physical parameters of a bubble domain material will be related to one another we need to have a quantitative description of the two most important atomic scale interactions in such materials. These are, first, the exchange energy, which describes the atomic scale interaction between neighbouring magnetic ions in a crystal and is responsible for the magnetic ordering of these ions; second, the anisotropy energy, which describes the interaction between these ordered magnetic ions and the crystal lattice and is responsible for the well defined easy direction of magnetisation. The exchange energy and the anisotropy energy are responsible for the domain wall energy, which came into our treatment of bubble domain statics in Chapter 2, § 2.3. Once we have a relationship between this wall energy and the more fundamental parameters of the magnetic material, we shall be able to define what kinds of magnetic crystal should be useful in bubble domain work.

3.1.1 *The exchange energy*

One of the most characteristic features of magnetic materials is the way in which the magnetisation of a saturated sample varies with temperature. Figure 3.1 shows some examples. If we plot the saturation magnetisation, M, normalised to its value at $0°K$, M_0, against temperature, we find that there is a critical temperature, T_c, at which the magnetisation collapses very rapidly. This temperature, T_c, is the Curie point.

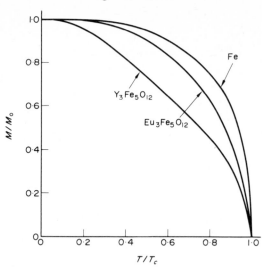

Figure 3.1. The variation of magnetisation with temperature in a simple ferromagnet, iron, and in two garnets in which the rare earth, or yttrium, plays no important part in the magnetisation.

As figure 3.1 shows, different materials may have very different behaviour in the way that the magnetisation varies with temperature below T_c but all show a sudden collapse of M at T_c. If we consider a particularly simple type of magnetic order, the ferromagnetic order of iron in which every atom in the crystal has the same magnetic moment, and at $0°K$, these are all aligned parallel to one another, it is clear that the collapse of M at T_c comes about because the thermal vibrations of the crystal lattice become sufficiently strong to uncouple the alignment between parallel atomic magnetic moments. This means that there must be a definite energy difference between the state of two neighbouring moments when they are (a) parallel, (b) anti-parallel and that this energy must be of the order of kT_c. It is this energy difference which is responsible for the magnetic order and it is referred to as the exchange energy (Chikazumi 1964), (Morrish 1965). The energy is $-E_{ex}$ when the neighbouring magnetic moments in iron are parallel and $+E_{ex}$ when they are anti-parallel. When $kT < E_{ex}$ the exchange interaction predominates in the crystal and the majority of the atomic magnetic moments drop into the lower energy state where they are parallel to one another (Kittel 1949).

The situation is very much more complicated in the garnets because there is not just one magnetic lattice, as in iron, but two or more. The garnets are cubic oxides, typified by yttrium iron garnet, $Y_3Fe_5O_{12}$, and a whole series of these compounds exist in which the yttrium in YIG may be replaced by one or more of the rare earths and the iron may be diluted by substituting trivalent ions like Ga^{3+} or Al^{3+}. There are eight formula units, that is $Y_{24}Fe_{40}O_{96}$, making up the unit magnetic cell in YIG (Lax and Button 1962) and we find that 16 of the 40 iron atoms occupy sites which are surrounded by six oxygen atoms in octahedral symmetry, while the other 24 iron atoms occupy sites surrounded by four oxygen atoms in tetrahedral symmetry. These sites are conventionally referred to as 'a' sites and 'd' sites, respectively. The yttrium occupies a completely different site, a 'c' site, which is surrounded by eight oxygen atoms in dodecahedral symmetry.

There is an exchange interaction in YIG which is very different to the simple one in pure iron because the atomic magnetic moments are separated by the oxygen atoms. This has the effect of reversing the sign of the exchange interaction so that anti-parallel ordering is the preferred, or $-E_{ex}$, state and the strongest of these interactions is between the 'a' sites and the 'd' sites. The exchange interaction between similar sites is, in contrast, positive, so that all the 'a' site iron atoms tend to have their magnetic moments aligned parallel to one another, all the 'd' site moments are parallel to one another but the 'a' lattice is anti-parallel to the 'd' lattice. This results in what is called a ferrimagnet, having two ferromagnetic sublattices anti-parallel to one another. The ratio of the number of atomic magnetic moments in one sublattice to the other is 24 to 16 so that there are 8 effective atomic magnetic moments per unit magnetic cell. This is the situation in YIG, where the yttrium is non-magnetic, and also in $Eu_3Fe_5O_{12}$, shown in figure 3.1, where the europium has a negligible magnetic moment. When there is a magnetic ion on the 'c' site, however, the garnet will show some very remarkable behaviour when the magnetisation is measured as a function of temperature. This is shown in figure 3.2 for holmium iron garnet.

When the atom occupying the 'c' site in the garnet structure has a magnetic moment, there is a negative exchange interaction between it and the atoms on the 'd' sites so that the 'c' site magnetic moments become aligned antiparallel to the 'd' site moments. In the case of holmium, the magnetic moment on the 'c' site is very large compared to that on the 'd' and 'a' sites so that the 'c' lattice will predominate magnetically at low temperatures where the alignment of these three magnetic sublattices is not strongly influenced by thermal fluctuations. Because the alignment of the 'c' lattice is due to a very indirect 'c'–'d'–'d'–'c' type of exchange interaction, the direct 'c' to 'c' site exchange being negligible, it follows that this 'c' lattice is very easily perturbed by thermal fluctuations and its net magnetic moment will drop rapidly with increasing temperature. As shown in figure 3.2, there will be a temperature at which the net magnetic moment of the 'c' lattice exactly compensates the net magnetic moment of the 'd' and 'a' lattices so that the magnetic moment of the whole crystal will vanish. The temperature at which this occurs is referred to as a compensation temperature and, to be exact, the magnetisation should be thought of as reversing sign at this

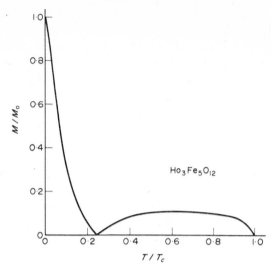

Figure 3.2. A garnet which has a magnetic ion on the rare earth sites will show a compensation temperature.

point. Experimentally, however, the measured magnetisation will always appear to have the same sign, as shown in figure 3.2, because a small external field will be applied during such measurements to keep the sample in the magnetically saturated state.

Even when we have such a complicated situation as we have in $Ho_3Fe_5O_{12}$, it is still true to say that the dominant exchange energy is of the order of kT_c. This is the result needed for a model of the domain wall and its energy and will enable us to relate the σ_w of Chapter 2 to some well defined parameters of the material. T_c is just such a parameter.

3.1.2   *The anisotropy energy*

The exchange energy, discussed above, causes the atomic magnetic moments in a magnetic crystal to align parallel or anti-parallel to one another but has very little to do with the actual direction which the aligned moments may take up relative to the crystallographic axes. In our consideration of magnetic bubble domain materials we can restrict the discussion to uniaxial crystals, that is those having only one easy or preferred direction of magnetisation. Because the magnetisation normally takes up this easy direction, it follows that there must be an energy associated with the direction of M, relative to the crystallographic axes, and that this energy is minimised when M takes up a particular direction. It is this energy which is termed the anisotropy energy.

In considering the exchange energy, we were able to relate its order of magnitude to the easily measurable parameter T_c. The same approach will be taken for the anisotropy energy. A clue to the required relation may be obtained when we consider an experimental measurement of uniaxial anisotropy which is shown in figure 3.3 where the magnetisation of a single crystal of cobalt is shown as a function of an externally applied magnetic

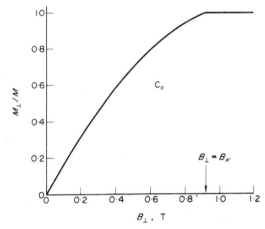

Figure 3.3 The magnetisation of a uniaxial material when a field, $B_\perp$, is applied perpendicular to the easy direction. Saturation will occur at the anisotropy field when the sample is a sphere.

field, $B_\perp$, which is perpendicular to the hexagonal axis of the crystal. Because cobalt has a very well defined easy direction of magnetisation parallel to the hexagonal axis, quite a large magnetic field, B_K, in figure 3.3, is needed to turn the magnetisation into the hexagonal plane. It is this field, B_K, which we use to characterise the anisotropy energy.

A measure of the anisotropy energy is simply the work done by the field, $B_\perp$, in moving the magnetic state of the sample from the origin of figure 3.3, where $M_\perp = 0$, to the point where $M_\perp = M$, the saturation magnetisation of the sample at this particular temperature. This is approximately equal to $MB_K/2$, because the characteristic shown in figure 3.3 is very nearly a straight line. To generalise this result we can say that, as the field required to turn the magnetisation to an angle ψ away from the easy axis will be

$$B_\perp = B_K \sin \psi,$$

the anisotropy energy density will be $M_\perp B_\perp/2$, or $(MB_K \sin^2 \psi)/2$ for a general angle ψ.

We have thus characterised the anisotropy energy by means of an empirical constant B_K, the anisotropy field, which may be measured on a sample of uniaxial material by taking a set of measurements of the kind shown in figure 3.3. The shape of the sample will have to be taken into account and this point will be discussed in § 3.3.5. The results of such measurements are usually given in the literature by quoting the anisotropy energy density

$$K_u = MB_K/2 \tag{3.1}$$

which was referred to in Chapter 2, § 2.4, in connection with the quality factor, Q, of a bubble domain material.

We are now in a position to justify the statement given in Chapter 2 concerning the need for a value of Q greater than unity. This can be seen from figure 3.4 where two possible situations in a given magnetic layer are shown. In figure 3.4(a), the magnetisation is normal to the plane of the layer while, in figure 3.4(b), the magnetisation lies in-plane. By using the

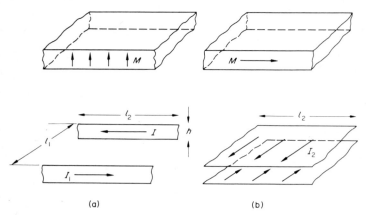

(a)   (b)

Figure 3.4. The magnetisation of a thin layer may be represented by equivalent currents.

equivalent current representation of the magnetisation, which was introduced in Chapter 2, we can see that the case of normal magnetisation is equivalent to two parallel currents, $I_1 = Mh$, while the case of in-plane magnetisation is equivalent to two current sheets, carrying a current $I_2 = M$ per unit length. The magnetic energy for the two cases may now be calculated from equation (2.14). In the case of figure 3.4(a), the equivalent inductance of the circuit is of the order of $\mu_0 l_2$, from equation (2.46) when the logarithmic term is taken as order unity, so that the magnetic energy for this case is

$$E_\uparrow \approx -\tfrac{1}{2}M^2 h^2 \mu_0 l_2/\pi \tag{3.2}$$

In the case of figure 3.4(b) the inductance is of the order $\mu_0 h l_1/l_2$ so that the magnetic energy $-\tfrac{1}{2}I^2 L$, is

$$E_\rightarrow = -\tfrac{1}{2}M^2 l_2^2 \mu_0 h l_1/l_2 \tag{3.3}$$

From equation (3.2) and (3.3), it follows that the energy density for normal magnetisation is $-\tfrac{1}{2}\pi\mu_0 M^2(h/l_1)$ while for in-plane magnetisation it is

$-\frac{1}{2}\mu_0 M^2$. When h/l_1 is very small, as it is in a thin layer, the in-plane magnetisation is a very much lower energy density situation and will be preferred if there are no other factors to be taken into account. When we include the effect of the anisotropy energy, however, it is clear that, if the uniaxial easy direction of magnetisation is normal to the plane of the layer, the situation shown in figure 3.4(a) will have zero anisotropy energy while that shown in figure 3.4(b) will have a positive anisotropy energy density equal to K_u, given by equation (3.1). It follows that the state of normal magnetisation will be preferred when

$$K_u > \tfrac{1}{2}\mu_0 M^2 \tag{3.4}$$

which corresponds to a value of Q greater than unity in equation (2.52). The better inequality (3.4) is satisfied, the more strongly will the required perpendicular magnetisation of the layer be preferred.

In conclusion, we have decided to represent the anisotropy energy by means of an experimentally determined parameter, either K_u or B_K. Unfortunately, both parameters are very temperature dependent so that this information will also be of great importance in considering a potential bubble domain material. The temperature dependence, and the quantum mechanical origin, of magnetocrystalline anisotropy have been a serious problem to theoretical physicists for a considerable time and a review of the early work is given by van Vleck (1945) and by Keffer and Oguchi (1960). Recent theoretical work is reviewed in a paper by Gyorgy *et al.* (1973) with particular reference to bubble domain materials and the growth induced anisotropy which we shall discuss in this chapter in § 3.2. An interesting example of the complexity in the temperature dependence of K_u in a bubble domain material has been given by Hagedorn *et al.* (1973) who measured K_u in $Y_2GdAl_{0.8}Fe_{4.2}O_{12}$. This material shows a compensation temperature, of the kind discussed here in connection with figure 3.2, and yet the anisotropy energy appears to fall smoothly as the temperature increases from $0°K$ to the Curie point. This means that B_K must rise to an immeasurably large value near the compensation temperature. The problem of measuring B_K in thin layers will be discussed in § 3.3.5.

3.1.3 *The domain wall energy*

We are now able to consider the origin of the magnetic domain wall energy, σ_w, which was introduced in Chapter 2 by means of figures 2.13 and 2.14. The first problem is to understand the stability of the domain wall itself and find out why it should take up a well defined width, δ, as shown in figure 2.13.

The simple Bloch wall, of the kind shown in figure 2.13, is one in which the atomic magnetic moments turn smoothly from pointing upwards, within the bubble domain, to downwards outside (Kittel 1949). This means that neighbouring moments will not be exactly parallel to one another but at a small angle, these small angular displacements adding up to $180°$ as we go from one side of the wall to the other. It follows that the exchange energy, discussed in § 3.1.1, will not be at its minimum value, $-E_{ex}$, between these atomic moments within the wall, but at the slightly higher value $-E_{ex} \cos \theta$, where θ is the small angle between neighbouring moments. This is an increase of $E_{ex}(1 - \cos \theta)$ for every pair of moments so that, if the number density of atomic moments is N, the increase in the exchange energy per unit area of domain wall will be

$$\Delta E_{ex} \approx N k T_c \theta^2 \delta/2 \qquad (3.5)$$

where we have substituted $k T_c$ for the exchange energy, from § 3.1.1, and put $(1 - \cos \theta) \approx \theta^2/2$ for small values of θ. If the neighbouring atomic moments are separated by a distance a, the angle $\theta = \pi a/\delta$ so that equation (3.5) becomes

$$\Delta E_{ex} = N k T_c \pi^2 a^2/2\delta \qquad (3.6)$$

and we see that the exchange energy will be reduced if the width of the wall, δ, increases because the angle, θ, between the neighbouring moments will be reduced.

The anisotropy energy, on the other hand, favours a very narrow domain wall because, then, the number of atomic moments which are at right angles to the easy direction of magnetisation will be minimised. Because the anisotropy energy varies as $\sin^2 \psi$, the angle ψ changing from zero to 180° as we go through the wall, the increase in the anisotropy energy per unit area of domain wall will be

$$\Delta E_a = K_u \delta/2 \qquad (3.7)$$

The total wall energy per unit area, the σ_w of Chapter 2, will thus be

$$\sigma_w = k T_c \pi^2/2a\delta + K_u \delta/2 \qquad (3.8)$$

which is the sum of equation (3.6), with N replaced by $1/a^3$, and equation (3.7). Differentiating equation (3.8) with respect to δ then shows that σ_w has a minimum value when

$$\delta = \pi(k T_c/a K_u)^{1/2} \qquad (3.9)$$

and substituting equation (3.9) into equation (3.8) gives

$$\sigma_w = \pi(k T_c K_u/a)^{1/2} \qquad (3.10)$$

as the wall energy per unit area in terms of our material parameters T_c and K_u. This is the result for the simple Bloch wall shown in figure 2.13. For the more complex wall, shown in figure 2.14, Malozemoff (1972) has shown that the wall energy is not significantly different until the bubble becomes so small that the wall width, δ, is a considerable fraction of the bubble diameter. The wall energy density then begins to increase and the contracting force, due to the wall energy, is reduced. This means that bubbles having a large number of segments of the kind shown in figure 2.14, will not collapse at the field we would expect, had σ_w remained constant, but at a considerably higher field. For this reason, bubble domains with a very large number of wall segments have become known as 'hard bubbles'. We shall see in Chapter 4 that there are good reasons for avoiding the generation of these hard bubbles in devices because their dynamic properties are inferior to bubble domains with the simpler wall structure. For this reason, there is no need to modify our results here, equations (3.9) and (3.10), particularly as these are only order of magnitude results intended as a guide in our choice of materials.

3.1.4 *The choice of a bubble domain material*

Referring back to Chapter 2, § 2.3.4, equation (2.38), we saw that the most significant parameter for a bubble domain material was its characteristic length, $\lambda = \sigma_w/\mu_0 M^2$. By means of equation (3.10) this may now be expressed as

$$\lambda = \frac{\pi}{\mu_0 M}\left(\frac{\mu_0 k T_c Q}{2a}\right)^{1/2} \qquad (3.11)$$

where we have substituted the quality factor, $Q = 2K_u/\mu_0 M^2$, which was discussed above and in Chapter 2, § 2.4. Equation (3.11) is a very important result because it shows that λ depends predominantly upon M and Q, T_c being more or less the same for all bubble domain materials if we exclude the possibility of using materials which are only magnetic at low temperatures. Table 3.1 shows a test of equation (3.11) which has been made by calculating the value of λ predicted by equation (3.11) and comparing this with the best measured value which can be found in the literature. The table shows the values of $\mu_0 M$, K_u and Q which have been measured or calculated from other measurements given in the references shown. The value of T_c is also shown. These values have then been substituted into equation (3.11) and the value of a, the atomic spacing of the elementary magnetic moments, taken as 10 Å, 10^{-9} m, in all cases. It can be seen from the last two columns of Table 3.1, that equation (3.11) certainly holds as an order of magnitude guide to the value of λ in a given material.

From an engineering, or device, point of view, the importance of equation (3.11) is best seen from figure 3.5 where equation (3.11) has been plotted as a function of $\mu_0 M$ for four values of Q, $T_c = 500°K$ and $a = 10^{-9}$ m. The measured values of λ and $\mu_0 M$, at 300°K, have been marked on figure 3.11 for the four materials described in Table 3.1.

Below the line $Q = 1$, on figure 3.5, bubble domains cannot exist because the preferred direction of magnetisation in the thin layer will be in-plane. Bearing this in mind, figure 3.5 shows that materials which support small bubbles must have a high value of $\mu_0 M$ provided we restrict ourselves to values of T_c which are well above room temperature. Consider $Eu_{0.6}Y_{2.4}$

Table 3.1.

Material	$\mu_0 M$ T	K_u J/m³	Q	T_c °K	λ (meas) μm.	λ (eqn 3.11) μm.	*Ref.*
BaFe₁₂O₁₉	0·400	$3\cdot3 \times 10^5$	5·2	720	$2\cdot12 \times 10^{-2}$	$4\cdot5 \times 10^{-2}$	Standley (1962) Kooy and Enz (1960)
YFeO₃	0·011	$4\cdot3 \times 10^4$	910	730	18·0	22·0	Rossol (1969)
Eu₀·₆Y₂·₄Ga₁·₁Fe₃·₉O₁₂	0·021	$\approx 10^3$	≈ 6	400	$\approx 0\cdot4$	$\approx 0\cdot7$	Gianola et al. (1969) Giess et al. (1971a)
Eu₂YFe₅O₁₂	0·149	3×10^4	3·4	550	$6\cdot9 \times 10^{-2}$	$8\cdot5 \times 10^{-2}$	Plaskett et al. (1972)

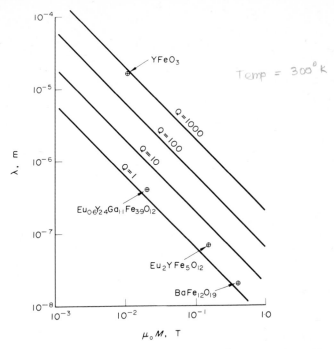

Figure 3.5. Equation (3.11) is plotted, for $T_c = 500°K$, $a = 10$ Å and four values of Q, as a function of $\mu_0 M$. This shows that small bubbles, needing small λ, will be found in materials with a high magnetisation.

$Ga_{1.1}Fe_{3.9}O_{12}$ as an example. The measured value of λ for this material is $0.4 \mu m$. So that, referring back to figure 2.22, we see that the smallest bubble diameter that we can work with in this material will be about $4 \mu m$. In order to make a device with smaller bubbles, and obtain a higher bit density, we must go to a higher value of $\mu_0 M$ and, consequently, tolerate higher drive fields to propagate the domains. It is just not possible to obtain smaller bubbles at the same low $\mu_0 M_s$ that is found in $Eu_{0.6}Y_{2.4}Ga_{1.1}$ $Fe_{3.9}O_{12}$. Equation (3.11) does show, however, that smaller bubbles combined with the advantage of low fields, could be obtained if we were prepared to build bubble domain devices which worked at low temperatures.

3.2   The Preparation of Magnetic Bubble Domain Materials

The single crystals required for bubble domain devices are prepared either as bulk crystals, grown from flux solutions or by the float zone technique, or as epitaxial layers which are grown in flux solutions or by chemical vapour deposition. A comprehensive treatment of crystal growth techniques may be found in the book by Laudise (1970) and the particular problem of growing oxide single crystals from flux solutions, the most important technique which will concern us here, has been the subject of a review by Elwell and Neate (1971). In this section a brief review of bubble domain materials will be given under the headings of the method of growth employed.

3.2.1   *Bulk crystals grown from flux solutions*

(i)   *The orthoferrites*   The first work on magnetic bubble domains (Bobeck 1967) was done using single crystal slices of the orthoferrites, of which $YFeO_3$, figure 3.5, is an example. There is a whole series of these compounds in which the yttrium of $YFeO_3$ is replaced by any rare earth or a mixture of rare earths, (Eibschutz *et al.* 1964), (Treves 1965).

The orthoferrites are weak ferromagnets, which means that they owe their magnetism to two sublattices which have the same magnetisation and are not quite anti-parallel to one another but canted at a small angle. The net magnetisation resulting from this is consequently at right angles to predominant direction of magnetic ordering. This is a very different situation to that found in the garnets, which was discussed in § 3.1.1, where the two, or more, sublattices were anti-parallel to one another and had different magnetisations so that the net magnetisation was in the same direction as the magnetic ordering.

The orthoferrites, as their name suggests, belong to the orthorhombic group of crystals which means that there are three orthogonal axes in the crystal along which the properties are all different. One of these axes is a very well defined easy direction of magnetisation so that, for bubble domain work, the crystal is cut and polished to provide thin slices with this easy axis normal to the plane of the slice. The in-plane properties are not isotropic, however, and the bubble domains which are observed in the orthoferrites are, in fact, elliptical (Della Torre and Dimyan 1970) and special precautions must be taken to overcome the confusing effects of the in-plane birefringence when such domains are observed using the Faraday effect (Tabor and Chen 1969).

Excellent single crystals of the orthoferrites may be grown from flux solutions and may already be in the form of thin platelets of the correct orientation (Bobeck *et al.* 1969) which are useful for experimental work. Otherwise, the bulk crystal must be orientated, sliced and polished and this is made a tedious and expensive business if the crystals are only a few millimetres in diameter as they often are, particularly as the size of the rare earth ion in the compound increases (Remeika and Kometami 1968). Large single crystals of $DyFeO_3$ and $YFeO_3$ have been prepared by a very different technique, the floating zone technique (Boyarchenkov *et al.* 1971), and this was considered a potentially good method for production on a commercial scale before the orthoferrites were abandoned as bubble domain device materials.

(ii) *The garnets* The orthoferrites may have been used to demonstrate the exciting possibilities of bubble domains in the first place, but they were never considered as suitable materials for practical device work because their low magnetisation and very high anisotropy brought them into the top left hand area of figure 3.5. This meant that the bubble domains in the orthoferrites were really large, over $100\,\mu m$ in diameter, and the high bit density which the bubble domain device promised could never be realised. The high anisotropy of the orthoferrites was also a considerable nuisance in experimental work because it was so large that, should the single crystal slice become magnetically saturated, it was not possible to nucleate domains in a slice without heating the sample up to a high temperature and cooling it down again.

Figure 3.5 shows the way which materials development had to be directed from the $YFeO_3$ point marked. This was towards much lower Q value, or lower anisotropy, and towards a slightly higher magnetisation. The bubble diameters would then be reduced to a few microns and the high bit density of the bubble domain device would be realised.

Development in this direction was first tackled by looking at mixed orthoferrites, such as $Sm_{0.55}Tb_{0.45}FeO_3$ (Rossol 1969), which have a much lower anisotropy energy than the simple orthoferrites like $YFeO_3$. Bubble domain diameters were reduced to below $10\,\mu m$ with these new materials but it is probably true to say that the real enthusiasm for the bubble domain device did not begin until Bobeck *et al.* (1971a) published their very unexpected discovery of bubble domains in garnet single crystals, cut from bulk flux grown crystals in such a way that there was sufficient growth induced anisotropy to provide a Q value greater than unity and take us to an area on figure 3.5, typified by the point marked $Eu_{0.6}Y_{2.4}Ga_{1.1}Fe_{3.9}O_{12}$. This growth induced anisotropy was not well understood at the time and was not easy to control; so that bulk crystals grown from flux solutions could be very similar in perfection and constitution and yet differ considerably in Q value. Attempts were made to overcome these problems by substituting cobalt into the garnets (van Uitert *et al.* 1971) and by using a high magnetostriction garnet which was deliberately stressed in order to produce an easy direction of magnetisation in the required direction (Giess *et al.* 1971b).

(iii) *The hexagonal ferrites* The final family of single crystals which may be prepared from flux solutions, and are potential bubble domain materials, are the hexagonal ferrites. These are typified by $BaFe_{12}O_{19}$, on figure 3.5, and have the high magnetisation and modest Q value needed for a real sub-micron sized bubble domain. The possibilities of these materials has been pointed out by Bobeck and Scovil (1971) in their very interesting review article, and Bobeck (1970) mentioned the possibilities of using hexagonal ferrites with lower magnetisation and larger bubble diameters, like

$PbAl_4Fe_8O_{19}$ and $BaAl_4Fe_8O_{19}$, but commented upon the poor dynamic properties of these materials. A comprehensive review of these interesting materials may be found in Smit and Wijn (1959).

3.2.2 *Liquid phase epitaxy*

The work on bulk crystals described above had shown that the garnets were the most useful materials for bubble domain device applications. The next development was to apply the technique of liquid phase epitaxy (LPE) to the problem of growing thin layers of bubble domain garnets.

LPE had already been applied to the problem of growing thin films of YIG for microwave devices by Linares (1968). The technique is very similar to the flux solution method of growing crystals (Laudise 1970), (White 1965) except that, instead of cooling the solution sufficiently to allow crystals to nucleate within the crucible, cooling is taken to only a few degrees below saturation and then a substrate crystal is dipped into the solution so that an epitaxial film of the solute crystal grows on the surface of the substrate.

In order to grow a thin magnetic layer of the correct orientation by LPE, we must have a non-magnetic substrate of the correct lattice spacing. A number of materials are available (Varnerin 1971) but the most readily available is gadolinium gallium garnet, $Gd_3Ga_5O_{12}$, and the vast majority of bubble domain LPE films have been grown on this substrate material. This explains, to a considerable extent, the rather complicated composition of the garnets which are used. For example, if yttrium iron garnet, $Y_3Fe_5O_{12}$, is prepared with some of the iron replaced by gallium we find that the gallium tends to prefer the smaller 'd' site (Lax and Button 1962), which was described above in § 3.1.1, and the net magnetisation of YIG will be reduced as more gallium is substituted until, at about 1·34 atoms of gallium per formula unit (Mee 1967), the magnetisation will be zero. From a magnetisation point of view, $Y_3Ga_{1.1}Fe_{3.9}O_{12}$ is a good bubble domain material, having $\mu_0M \approx 0.02$ T, and showing sufficient growth induced anisotropy to give a Q value above unity. This simple gallium substituted YIG does not have the correct crystal lattice spacing to grow epitaxially upon $Gd_3Ga_5O_{12}$, however, so that our next step is to modify the chemistry of the rare earth site so that we can get a good lattice match between the substrate and the magnetic layer and grow the required orientation for our bubble domain film. This can be done without changing the magnetic properties too much provided we use yttrium and europium on the rare earth site, both being virtually non-magnetic.

The literature on liquid phase epitaxy, even when we restrict ourselves to bubble domain materials, is considerable and is growing rapidly. Recent papers by Giess *et al.* (1972) and Ghez and Giess (1973) describe the technique in detail and give further references. The LPE process certainly appears to be very successful and can be applied on a production basis (Hewitt *et al.* 1973). The garnet film which was used to prepare figures 1.5, 1.6, 1.8 and 1.9 was made using this technique.

3.2.3   *Epitaxial films by chemical vapour deposition*

The growth of epitaxial garnet films by chemical vapour deposition (CVD) for bubble domain devices has been developed with reasonable success (Mee *et al.* 1971). In this process, the metals of the garnet are carried to the single crystal substrate, upon which epitaxial growth is planned, in the form of halide vapours. The garnet is formed at the substrate by a reaction of these halides with water vapour. In principle this technique should produce higher quality films than the LPE process and should be easier to control. It turns out, however, to be full of difficulties and very much more work is needed to set the apparatus up. For these reasons, CVD has almost been abandoned as a technique for preparing bubble domain films.

One fundamental difference between the CVD and LPE processes should be mentioned in conclusion. This concerns the nature of the magnetic anisotropy which is predominantly growth induced in LPE films and is also influenced by the small amount of lead or bismuth which is always found in these films and comes from the flux which is used as a solvent. The CVD films contain no such impurity and their anisotropy appears to be controlled by stress in the film which comes about because of a deliberate mismatch between the substrate and the epitaxial layer. A similar stress induced anisotropy may play some part in the LPE films too (Giess *et al.* 1971b).

3.2.4   *Established bubble domain materials*

In conclusion, what can be said about the materials which have really been used in the bubble domain devices which have been made so far? If we exclude the early work using orthoferrites, it would appear that two kinds of garnet, both prepared by LPE, have been favoured—the first group is based on $(EuY)_3(GaFe)_5O_{12}$, of which the garnet marked on figure 3.5, which has $Eu_{0.6}$ and $Ga_{1.1}$, is typical. Sadagopan *et al.* (1971) constructed a 400 bit shift register on a very similar garnet film $(Eu_{0.5}, Ga_{1.0})$ using 4 μm diameter bubbles and this worked at 100 KHz. Almasi *et al.* (1971) tested encoding and decoding overlay patterns for a proposed mass memory on a garnet film with $Eu_{0.7}Ga_{1.2}$. The film thickness was 15 μm and the bubble diameter 12 μm.

The first group of garnets are characterised by the fact that the rare earth ions play very little part in the magnetic behaviour, being present only to modify the lattice so that it fits the substrate. Going from $Eu_{1.5}Ga_{1.0}$ to $Eu_{0.7}Ga_{1.2}$ does change the value of μ_0M from about 0·025 T to 0·01 T, and so changes the bubble diameter considerably, the value of Q being about 3 in both cases. The material is thus a very sensible one to work with although its dynamic properties, which will be considered in Chapter 4, do leave something to be desired.

The second group of garnets which have been used in bubble domain devices are of the form $(ErEu)_3(GaFe)_5O_{12}$, $(ErGd)_3(GaFe)_5O_{12}$ (Bobeck *et al.* 1971b), $(EuGd)_3(GaFe)_5O_{12}$ (Argyle *et al.* 1971) and $(ErEu)_3(GaAlFe)_5$

O_{12}, (Le Craw and Pierce 1971). These garnets are of the kind which show
a compensation temperature, a phenomenon which was discussed in § 3.1.1
in connection with figure 3.2, and the object is to arrange that the compensation temperature and the Curie temperature occur more or less equally
on either side of the temperature at which the bubble domain device will
be used. The flat part of the characteristic shown in figure 3.2 will then
occur at the working temperature and we would hope that the device could
work over quite a wide temperature range. Further modifications to the
rare earth chemistry of these garnets could then be made to introduce some
temperature compensation which would be advantageous. A flexibility in
the rare earths can also be used to modify the dynamic properties, as we
shall see in Chapter 4.

3.3 Measurement of Bubble Domain Material Properties

3.3.1   *The parameters to be measured*

The parameters of real interest in a bubble domain film are the range of
stable bubble diameters which can be maintained between collapse and
run-out, the range of bias fields needed for these diameters, the thickness
of the film and then the dynamic properties of these bubble domains.
Dynamic properties will be considered in Chapter 4.

   The mere observation of bubble domains in a thin layer does not mean
that the material is of any use for device work. For example, bubble domains
can be seen in thin single crystals of cobalt (Grundy *et al.* 1971) which has
a Q value below unity. Isolated bubble domains would be rather unstable
in such a material. De Bonte (1971) has shown that bubbles can persist at
Q values as low as 0·75.

   Referring to figure 2.21, we see that, if we know the thickness of the magnetic layer, a measurement of the run-out and collapse diameters will enable
us to estimate λ/h and then go to figure 2.20 and relate the collapse and
run-out fields to $\mu_0 M$. This will give us an estimate of the important material
parameters, λ, h and $\mu_0 M$. An example will show how useful an estimate
this might be.

3.3.2   *An example*

Bubble domains were generated in an LPE film of $(EuY)_3(GaFe)_5O_{12}$ by the
method described in Chapter 1, in connection with figure 1.9. By carefully
moving the small coil which generates the pulsed magnetic field and by
manipulating the bias field, it was possible to obtain a few virtually isolated
bubble domains within the field of view and careful measurements of their
run-out and collapse fields and diameters could be made.

   The run-out diameter was found to be between 12·0 μm and 13·0 μm at a
field of $46\cdot5 \times 10^{-4}$ T while the collapse diameter was between 4·5 μm and
5·2 μm at a field of $60\cdot5 \times 10^{-4}$ T. The fields were very well defined and

known with high accuracy and the spread in diameters, over several rep-
etitions of the experiment, were real and increased by the difficulty of
measuring such small diameters with a simple microscope.

The thickness of this particular film, measured by infrared interference
in the manner described in § 3.3.3, was 6·25 μm so that the results given above
can be put on to figure 2.21 to see how well they fit the theory given in
Chapter 2. Figure 3.6 shows the result. Figure 3.6(a) shows the bottom left

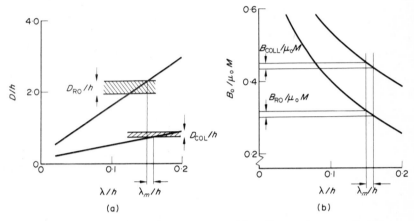

Figure 3.6. Fitting experimental measurements of the collapse and run-out diameters
to the theory of Chapter 2 so that λ and M may be estimated.

hand corner of figure 2.21 on a larger scale and we can see that the experimental results only make sense if we assume that our measurements of
collapse diameter are too large while those of run-out diameter are too small.
This would be the case because we cannot really measure the collapse
diameter; instead we measure the slightly larger diameter just before collapse. Similarly the run-out of the bubble domain will be precipitated in
that any small imperfection in the film will trigger off the run-out process.

It would thus appear sensible to use the lower bound on D_{COL}/h and the
upper bound on D_{RO}/h and these are shown on figure 3.6(a) giving the range
λ_m/h. When this is transferred to figure 3.6(b), which is a scaled up copy of
our previous figure 2.20, we find that our measurement of B_{COL}, which we
expect to be low, predicts a value of $\mu_0 M$ between 135×10^{-4} and
139×10^{-4} T; while our measurement of B_{RO}, which we expect to be high,
predicts a value of $\mu_0 M$ between 155×10^{-4} and 147×10^{-4} T. As a result
of this we may conclude that our measurements do fit the simple theory
and that $\mu_0 M$ lies between these values at $(143 \pm 5) \times 10^{-4}$ T. From our
value of λ_m/h, the value of λ for this film is 0.96 ± 0.04 μm. We have made
no measurements of the anisotropy but our equation (3.11) predicts that
$Q = 5.4$ which is also a very sensible result.

Measurements of the kind described above, which are based upon bubble domain measurements, are obviously very useful in deciding whether a given magnetic layer is suitable for device work and for telling us what value of bias field would be needed. It is important, however, to have more fundamental measurements of the magnetic properties and, in the following sections, we shall briefly review the methods which have been used for characterising the static properties of bubble domain films.

3.3.3 *Thickness measurements*

One of the most important parameters of a bubble domain layer is its thickness, h. The easiest way of measuring this is interferometrically and an example of the kind of data which such a measurement provides is shown in figure 3.7. Most of the garnet layers used for bubble domain work are

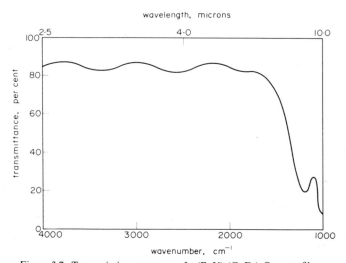

Figure 3.7. Transmission spectrum of a $(EuY)_3(GaFe)_5O_{12}$ LPE film.

transparent from the green, $0.5\ \mu m$ wavelength, up to a wavelength of about $10\ \mu m$ where lattice vibrations begin to dominate the absorption of radiation. Absorption at the shorter wavelengths is electronic and is closely associated with the atomic processes responsible for the Faraday effect which we use to view the domains. For this reason, Faraday rotation micrographs, such as those shown in figures 1.6 to 1.9, are taken using green or yellow light to get good contrast.

At the longer wavelengths, shown in figure 3.7, the garnet will be quite transparent and show very little dispersion so that we may interpret the fluctuations which are observed in the transmittance as being due to interference between rays which are internally reflected between the air-layer interface and the layer-substrate interface. On a linear wavenumber scale, the maxima of the transmittance plot will be equally spaced, $(v_1 - v_2)$ $= (v_2 - v_3)$ etc., and the thickness of the layer will be given by

$$h = \frac{1}{2(v_1 - v_2)n} \qquad (3.12)$$

(Born and Wolf 1964), where n is the refractive index of the layer, which we assume to be constant.

The infrared refractive index of magnetic garnets has been studied by Johnson and Walton (1965) and their results would lead us to expect a value of infrared refractive index near 2·2 for the garnets used in bubble domain work. LPE layers may contain 1 to 3 per cent of lead, however (Giess *et al.* 1971c) and this increases the refractive index to nearer 2·5 for most LPE garnets. A further complication is that the interference method appears to measure the thickness between the top of the epitaxial layer and the bottom of the lead diffusion into the substrate. This adds a typical 0·1 μm to our thickness measurement anyway (Ghez and Giess 1973).

For this reason, it is essential to measure the thickness of epitaxial layers by a more accurate method so that the true value of the refractive index can be found. This can be done by bevelling the layer at a shallow angle, measuring this angle very accurately and measuring the width of the bevelled layer as it shows up against the polished background. The technique used by Ghez and Giess (1973) to determine thickness really accurately was to crack one of their layers and view it on edge in the scanning electron microscope. A further accurate method is to grind a spherical depression into the layer.

3.3.4 *Determining the composition*

It is very difficult to determine the exact chemical composition of a bubble domain layer and, for this reason, it should be noted that the majority of authors give only nominal compositions. Crystals which grow in flux solutions, both bulk crystals and LPE layers, do not have the composition of the material which was originally dissolved and in the chemical vapour deposition process one may know accurately what proportions of vapour enter the system, but have very little idea what proportions reach the reaction zone. An interesting attempt to overcome these problems in the LPE process has been described by Robbins *et al.* (1971) in which the required garnet composition is made as a coprecipitate and epitaxial growth takes place in a very small amount of flux.

It is also important to know the composition if the properties are to be understood on a fundamental level. Analytical techniques which simply give the chemical formula are only a partial answer to the problem because, in a material like $(EuY)_3(FeGa)_5O_{12}$ for example, the gallium may prefer the 'd' site as stated in § 3.2.2, but it is not exclusive in this preference and gallium may be found on the 'a' sites after heat treatment. Cronemeyer *et al.*

(1971) have used spectrophotometric analysis to determine the ferric ion concentration on the 'a' and 'd' sites in epitaxial layers of bubble domain garnets.

3.3.5   *Accurate measurement of the static magnetic properties*

The important magnetic parameters are the magnetisation, M, the characteristic length, λ, and the anisotropy energy, K_u. The temperature dependence of all these parameters should be known and this will include the Curie temperature and the value of the compensation temperature if the material has one.

It must always be remembered that these measurements have to be carried out on very thin layers attached to substrates which are perhaps 0·5 mm thick. The substrate is strongly paramagnetic, because it contains gadolinium, and this severely reduces the accuracy of conventional magnetometer measurements. For this reason, and because of the importance of knowing λ accurately, a number of techniques have been worked out which rely upon observing the periodicity of the unmagnetised domain pattern in a bubble domain film either as a strip pattern or as a bubble domain array. The information which can be obtained from a strip pattern was first discussed by Kooy and Enz (1960) and has been applied to bubble domain materials by Cape and Lehman (1971), Craik and Cooper (1972), De Jonge and Druyvesteyn (1971) and Fowlis and Copeland (1971). These authors agree that these methods give an accuracy of about 5 per cent in the value of λ and 10 per cent in the value of M. In the author's experience, however, these methods appear to be rather inaccurate because they rely upon observing the minimum energy domain pattern which is calculated for an infinite layer. The domain pattern in a layer of finite size depends mainly upon the number of domains which happen to be nucleated and these will certainly arrange themselves in a minimum energy configuration but it is the minimum energy configuration for that particular number of domains. The minimum energy configuration which is assumed by the authors mentioned above may not be attained until either more domains are nucleated or some existing domains removed and there is no way of telling what the situation really is. A further difficulty with this method is that it assumes a simple Bloch wall is always involved whereas a more complex wall is far more likely and such a wall may have a slightly different energy each time the domain pattern is produced.

The measurement of the uniaxial anisotropy of a bubble domain layer may be made with high accuracy using ferromagnetic resonance techniques (Le Craw and Pierce 1971) if the resonance line width is reasonably narrow. Unfortunately, from the point of view of this method, bubble domain materials with good dynamic properties may have very broad line widths, as we shall see in Chapter 4. A less accurate method is to measure the field required to pull the magnetisation of the bubble domain layer in-plane.

This field is not equal to the anisotropy field, B_K, which was introduced in connection with figure 3.3, because we are dealing with a thin layer sample. Reference to equations (3.2) and (3.3) show that, if the field required to magnetise a thin layer in-plane is B_{KA}, we have

$$B_{KA} = B_K - \mu_0 M + \mu_0 M(h/l_1) \tag{3.13}$$

where the last term may be neglected in any practical bubble domain layer. It follows from equations (2.52) and (3.1) that the quality factor of the layer is

$$Q = 1 + \frac{B_{KA}}{\mu_0 M} \tag{3.14}$$

Experimentally, a measurement of B_{KA} may be made in a Faraday rotation hysteresis loop plotter, which is simply the arrangement shown in figure 1.3 with the microscope replaced by a photodetector and the polariser and analyser at 45°, instead of being nearly crossed as they are when Faraday rotation micrographs are taken. For anisotropy measurements, the sample is placed in a pulsed transverse field which is increased in amplitude until the signal from the photodetector indicates in-plane saturation. More accurate techniques involving the same principle have been described by Josephs (1972) in his excellent review paper on magnetic measurements applied to bubble domain materials. Josephs also discusses the importance of magneto-optical methods in the measurement of magnetisation and coercive force.

The coercive force of bubble domain layers is an important parameter and should be extremely small, perhaps less than 10^{-5} T (Craik *et al.* (1973)). It is not easy to measure such small coercive forces and some very sensitive methods have been proposed which involve exploiting the repulsive force between bubble domains and should lead to a true bubble coercive force, as opposed to a wall coercive force. Thiele (1971) mentions this method and gives a result which follows from his general treatment of the force on a magnetic bubble domain in a field gradient. This is a result we shall need in Chapter 4, so that it will be useful to look at it here and obtain the formula needed for the calculation of coercive force.

Figure 3.8 shows the bubble domain, with its effective wall current $2Mh$, in a gradient field which is equal to the required bias field, B_0, at the centre of the bubble. We should expect the bubble to move to the right, as we discussed in connection with figure 1.11. The force on the bubble will be a simple modification of equation (2.13). Normal to the wall, at a point such as P in figure 3.8, the force will be

$$\Delta F_0 = \left(B_0 + \frac{\partial B_z}{\partial x} R \cos \theta - B_c \right)(2Mh)(R \, d\theta) \tag{3.15}$$

where the coercive force of the material is represented by B_c and has the

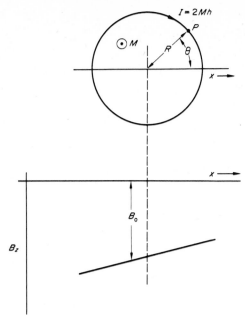

Figure 3.8. The force on a magnetic bubble in a bias field gradient.

negative sign for $-\pi/2 < \theta < +\pi/2$ when we assume that it is a simple wall coercivity which subtracts from the normal force on the wall. On the left hand side of the bubble domain, equation (3.15) also applies but with $-B_c$ replaced by $+B_c$. Assuming that the bubble remains circular, and at the radius it would have in a uniform field, the total force to the right will be

$$F_x = 2Mh\left\{\int_{-\pi/2}^{+\pi/2} [B_0 + (\partial B_z/\partial x)R\cos\theta - B_c]R\cos\theta\, d\theta \right.$$

$$\left. + \int_{+\pi/2}^{+3\pi/2} [B_0 + (\partial B_z/\partial x)R\cos\theta + B_c]R\cos\theta\, d\theta \right\} \quad (3.16)$$

and, when we take $\partial B_z/\partial x$ to be a constant, this becomes

$$F_x = 2Mh[\pi R(\partial B_z/\partial x) - 4B_c]R \quad (3.17)$$

It follows that the bubble domain will only move when the applied field gradient exceeds $4B_c/\pi R$. This may be used as a measure of coercive force by applying a known field gradient. In the case of two isolated neighbouring bubbles, we may say that the field gradient is due to the dipole field of one of the bubbles acting on the other. This dipole field is approximately

$$B_z = -(\mu_0/4\pi)(2Mh)(\pi R^2)(1/x^3) \quad (3.18)$$

(Feynman (1964)), when the separation is several radii. Differentiating equation (3.18) with respect to x, and equating the result to $4B_c/\pi R$, then gives

$$B_c = (\mu_0 M)\frac{3\pi R^3 h}{8x^4} \tag{3.19}$$

which is Thiele's (1971) result for the coercive force in a material when two isolated bubbles are observed to be separated by a distance x between their centres. This is clearly not a very useful result when the bubble domains are only a few microns in diameter because of the difficulty in measuring R and x accurately.

References

Almasi, G. S., Canavello, B. J., Giess, E. A., Hendel, R. J., Horstmann, R. E., Jamba, T. F., Keefe, G. E., Powers, J. V., and Rosier, L. L., 1971, *AIP Conf. Proc.*, No. 5, p. 220.

Argyle, B. E., Slonczewski, J. C., and Mayadas, A. F., 1971, *AIP Conf. Proc.*, No. 5, p. 175.

Bobeck, A. H., 1967, *Bell Syst. Tech. J.*, **46**, 1901.

Bobeck, A. H., Fischer, R. F., Perneski, A. J., Remeika, J. P., and van Uitert, L. G., 1969, *IEEE Trans. on Magnetics*, MAG 5, 544.

Bobeck, A. H., 1970, *IEEE Trans. on Magnetics*, MAG 6, 445.

Bobeck, A. H., Smith, D. H., Spencer, E. G., van Uitert, L. G., and Walters, E. M., 1971a, *IEEE Trans. on Magnetics*, MAG 7, 461.

Bobeck, A. H., Fischer, R. F., and Smith, J. L., 1971b, *AIP Conf. Proc.*, No. 5, p. 45.

Bobeck, A. H., and Scovil, H. E. D., 1971, Magnetic Bubbles, *Scientific American*, June 1971, p. 78.

Born, M., and Wolf, E., 1964, *Principles of Optics*, 3rd edition. Pergamon Press, Oxford, p. 62.

Boyarchenkov, M. A., Raev, V. K., Samarin, M. I., Balbashov, A. M., and Chervonenkis, A. Ya., 1971, *IEEE Trans. on Magnetics*, MAG 7, 352.

Cape, J. A., and Lehman, G. W., 1971, *J. appl. Phys.*, **42**, 5732.

Chikazumi, S., 1964, *Physics of Magnetism*. Wiley, New York, **138**, p. 188.

Craik, D. J., and Cooper, P. V., 1972, *J. Phys. D.*, **5**, p. L37.

Craik, D. J., Cooper, P. V., and Myers, G., 1973, *J. Phys. D.*, **6**, 872.

Cronemeyer, D. C., Giess, E. A., Klokholm, E., Argyle, B. E., and Plaskett, T. S., 1971. *AIP Conf. Proc.*, No. 5, p. 115.

De Bonte, W. J., 1971, *AIP Conf. Proc.*, No. 5, p. 140.

De Jonge, F. A., and Druyvesteyn, W. F., 1971, *AIP Conf. Proc.*, No. 5, p. 130.

Della Torre, E., and Dimyan, M. Y., 1970, *IEEE Trans. on Magnetics*, MAG 6, 489.

Eibschutz, M., Gorodetsky, G., Shtrikman, S., and Treves, D., 1964, *J. appl. Phys.*, **35**, 1071.

Elwell, D., and Neate, B. W., 1971, *J. Mat. Sci.*, **6**, 1499.

Feynman, R. P., Leighton, R. B., and Sands, M., 1964, *Lectures on Physics*, vol. II, p. 14–7. Addison-Wesley Publishing Co. Inc., Reading, Mass.

Fowlis, D. C., and Copeland, J. A., 1971, *AIP Conf. Proc.*, No. 5, p. 240.

Gianola, U. F., Smith, D. H., Thiele, A. A., and van Uitert, L. G., 1969, *IEEE Trans. on Magnetics*, MAG 5, 558.

Giess, E. A., Argyle, B. E., Cronemeyer, D. C., Klokholm, E., McGuire, T. R., O'Kane, D. F., Plaskett, T. S., and Sadagopan, V., 1971a, *AIP Conf. Proc.*, No. 5, p. 110.

Giess, E. A., Calhoun, B. A., Klokholm, E., McGuire, T. R., and Rosier, L. L., 1971b, *Mat. Res. Bull.*, **6**, 317.

Giess, E. A., Argyle, B. E., Calhoun, B. A., Cronemeyer, D. C., Klokholm, E., McGuire, T. R., and Plaskett, T. S., 1971c, *Mat. Res. Bull.*, **6**, 1141.

Giess, E. A., Kuptsis, J. D., and White, E. A. D., 1972, *J. Cryst. Growth*, **16**, 36.

Ghez, R., and Giess, E. A., 1973, *Mat. Res. Bull.*, **8**, 31.

Grundy, P. J., Hothersall, D. C., Jones, G. A., Middleton, B. K., and Tebble, R. S., 1971, *AIP Conf. Proc.*, No. 5, p. 155.

Gyorgy, E. M., Sturge, M. D., van Uitert, L. G., Heilner, E. J., and Grodkiewicz, W. H., 1973, *J. appl. Phys.*, **44**, 438.

Hagedorn, F. B., Tabor, W. J., and van Uitert, L. G., 1973, *J. appl. Phys.*, **44**, 432.

Hewitt, B. S., Pierce, R. D., Blank, S. L., and Knight, S., 1973, *IEEE Trans. on Magnetics*, MAG, **9**, 366.

Johnson, B., and Walton, A. K., 1965, *Brit. J. appl. Phys.*, **16**, 475.

Josephs, R. M., 1972, Characterisation of the magnetic behaviour of bubble domains, *18th Conf. on Magnetism and Magnetic Materials*, Denver, Nov. 1972 (*AIP Conf. Proc.* No. 10, p. 286).

Keffer, F., and Oguchi, T., 1960, *Phys. Rev.*, **117**, 718.

Kittel, C., 1949, *Rev. mod. Phys.*, **21**, 541.

Kooy, C., and Enz, U., 1960. *Philips Res. Repts.*, **15**, 7

Laudise, R. A., 1970, *The Growth of Single Crystals*. Prentice Hall Inc., Englewood Cliffs, N.J., p. 293.

Lax, B., and Button, K. J., 1962, *Microwave Ferrites and Ferrimagnets*. McGraw-Hill Publishing Co., New York, pp. 125–135.

LeCraw, R. C., and Pierce, R. D., 1971, *AIP Conf. Proc.*, No. 5, p. 200.

Linares, R. C., 1968, *J. Cryst. Growth*, **34**, 443.

Malozemoff, A. P., 1972, *App. Phys. Lett.*, **21**, 149.

Mee, C. D., 1967, *Contemp. Phys.*, **8**, 385.

Mee, J. E., Pulliam, G. R., Heinz, D. M., Owens, J. M., and Besser, P. J., 1971, *App. Phys. Lett.*, **18**, 60.

Morrish, A. H., 1965, *The Physical Principals of Magnetism*. Wiley, New York, p. 370.

Plaskett, T. S., Klokholm, E., Hu, H. L., and O'Kane, D. F., 1972, Magnetic bubble domains in $(EuY)_3Fe_5O_{12}$ films on $Sm_3Ga_5O_{12}$ substrates, *18th Conf. on Magnetism and Magnetic Materials*, Denver, Nov. 1972 (*AIP Conf. Proc.* No. 10, p. 319).

Remeika, J. P., and Kometami, T. Y., 1968, *Mat. Res. Bull.*, **3**, 895.

Robbins, M., Licht, S. J., and Levinstein, H. J., 1971, *AIP Conf. Proc.*, No. 5, p. 101.

Rossol, F. C., 1969, *IEEE Trans. on Magnetics*, MAG 5, 562.

Sadagopan, V., Hatzakis, K. Y., Ahn, K. Y., Plaskett, T. S., and Rosier, L. L., 1971, *AIP Conf. Proc.*, No. 5, p. 215.

Smit, J., and Wijn, H. D. J., 1959, *Ferrites*. Wiley Inc., New York.

Standley, K. J., 1962, *Oxide Magnetic Materials*. Oxford University Press, London.

Tabor, W. J., and Chen, F. S., 1969, *J. appl. Phys.*, **40**, 2760.

Thiele, A. A., 1971, *Bell Syst. Tech. J.*, **50**, 725.

Treves, D., 1965, *J. appl. Phys.*, **36**, 1033.

van Uitert, L. G., Gyorgy, E. M., Bonner, W. A., Grodkiewicz, W. H., Heilner, E. J., and Zydzik, G. J., 1971, *Mat. Res. Bull.*, **6**, 1185.

van Vleck, J. H., 1945, *Rev. mod. Phys.*, **17**, 27.

Varnerin, L. J., 1971, *IEEE Transactions on Magnetics*, MAG 7, 404.

White, E. A. D., 1965, *Techniques of Inorganic Chemistry* (Jonassen, ed.), Vol. IV, Interscience, New York, pp. 31–64.

4   Bubble Domain Dynamics

4.1   Ferromagnetodynamics

The dynamics of ferromagnetic materials has been one of the most fascinating problems in applied magnetism for some time and has a considerable literature. The earliest treatment of the problem of a moving magnetic domain wall may be found in a remarkable paper 'On the theory of the dispersion of magnetic permeability in ferromagnetic bodies' by Landau and Lifshitz (1935). This paper was published in English originally and, in the space of only sixteen pages, gave the first theoretical explanation for the magnetic domain patterns which had been observed by Bitter (1931) a few years previously. The stability and thickness of the domain walls was also derived and the dynamics of wall motion in an applied magnetic field was considered for the first time. The paper concluded with a treatment of what we now call ferromagnetic resonance.

After this paper by Landau and Lifshitz, little was heard about dynamic problems in magnetism until around 1950. There was then an explosion of interest, brought about by the introduction of the ferrite core store into computers and by the hope of using thin ferromagnetic films as high-speed data storage elements. This early experimental and theoretical work has been reviewed by Kittel and Galt (1956) and by Gyorgy (1960, 1963).

The introduction of the bubble domain in 1967 precipitated a further surge of interest in dynamic problems. The motion of a bubble domain within a single crystal layer of magnetic material is, in many ways, a more well defined problem than the one of simple magnetisation reversal; the problem which had occupied the attention of theoreticians previously. In fact, bubble domain dynamics has turned out to be a problem which is not only of crucial importance technically, for the realisation of useful devices, but also a problem of great interest from the fundamental point of view.

Our problem here is to work out a theoretical model of bubble domain dynamics which is sufficiently general, and yet simple enough, to be useful in the design of devices. The questions we must be able to answer concern the shape and intensity of the magnetic fields needed for bubble domain propagation; the kind of fields which we discussed qualitatively in § 1.4 of Chapter 1. This problem is very closely related to the design of the magnetic overlay, which produces the propagating field, and a great deal of the work discussed in this chapter will lead directly into Chapter 5 where the overlay problem will be considered.

In order to choose a theoretical model, we shall first review what is known experimentally about bubble domain dynamics and propagation. We shall see that we are not only concerned with the atomic scale processes which must occur within the moving domain walls themselves, the processes which control the way in which the atomic scale magnetic moments actually rotate as the wall moves, but that we must also consider the stability of the bubble domain in motion and the way in which it may change size and shape.

4.2 Experimental Work on Bubble Domain Dynamics

4.2.1 *The concept of wall mobility*

The paper by Landau and Lifshitz (1935), which was referred to above, predicted that the velocity of a magnetic domain wall would be proportional to the local field acting upon it. Considered in terms of the equivalent current model of domain walls, which was introduced in figure 2.4, the domain wall in a bubble domain layer of thickness h is equivalent to a current $2Mh$ and the force acting upon it will be $(2Mh)B$, per unit length. The local field, B, is simply the sum of all the externally applied fields and the fields produced by other, neighbouring, domain walls in the particularly simple case of a straight infinite wall, of the kind shown in figure 2.4. A velocity which is proportional to this force then implies that there is a viscous force, proportional to velocity, acting upon the moving wall and if this viscous force is βv_n, per unit area of wall, equilibrium applies when, per element of domain wall length, we have

$$\beta v_n h = (2Mh)B \tag{4.1}$$

so that for constant magnetisation, M, and constant β we have a velocity normal to the wall

$$v_n = \mu_w B \tag{4.2}$$

where $\mu_w = 2M/\beta$ is referred to as the wall mobility. The viscous force acting upon an element of wall length dl may now be expressed as

$$F_n = \beta v_n h \, dl = \frac{2Mh}{\mu_w} v_n \, dl \tag{4.3}$$

Landau and Lifshitz (1935) had been able to show how the viscous parameter, β, was related to the atomic scale processes occurring in the simple Bloch wall as it moved. The simple linear relationship between applied field and velocity, equation (4.2), had been found experimentally in ferrites at velocities of a few metres per second by Galt (1954), who also made measurements at much higher velocities and found that there was a limit to the domain wall velocity in nickel zinc ferrite at around 650 m/s. This is a very high velocity from a bubble domain device point of view so that it was sensible to use the concept of a simple wall mobility for the early work on bubble dynamics (Bobeck *et al.* 1969 and Perneski 1969) as non-linear theories of domain wall motion were not yet established (Palmer and Willoughby 1967).

The orthoferrites were the first bubble domain materials and Rossol (1969) made measurements of the wall mobility in a whole series of these materials by observing the response of straight domain walls to a small oscillating magnetic field which was directed normal to the plane of the single crystal layer. The amplitude of the wall oscillations was measured optically by observing their motion stroboscopically with light modulated at radio frequencies. In $YFeO_3$, for example, a value of $\mu_w = 10^5$ m/s per T was measured at room temperature and there was no sign of any non-linearity at the velocities expected in practical devices.

4.2.2 Bubble domain motion in a bias field gradient

Rossol and Thiele (1970) set up an experiment to see if the simple wall mobility concept could be applied to a bubble domain in translation. The essential details of this experiment are shown in figure 4.1. An isolated bubble domain was located in a bias field minimum which was produced by means of a saturated permalloy disc close to the surface of a single crystal slice of $Sm_{0.55}Tb_{0.45}FeO_3$ orthoferrite. Figure 4.1 shows how the permalloy disc, magnetically saturated by the in-plane drive field B_D, produces a minimum in the bias field by acting against the field B_0. There is a radial field minimum and a circumferential field minimum so that, at the correct value of B_0, a stable bubble domain will be held near the edge of the permalloy disc. If the in-plane drive field is now rotated, the bubble domain should rotate as well, lagging behind the field by a small constant angle so that it sees a local field at its wall in order to cause its motion.

Rossol and Thiele (1970) observed this rotation stroboscopically and measured the angle by which the bubble domain lagged behind the direction of the drive field as a function of frequency. The complicated field problem, illustrated by figure 4.1, was analysed numerically but the model used was fundamentally one later published by Thiele (1971) and used the simple wall mobility concept in that the viscous drag force, given by equation (4.3), was considered to act around the bubble domain wall. The situation is that shown in figure 4.2 where a translational velocity v_x implies a velocity normal to the wall of $v_x \cos \theta$ and a total viscous drag force

$$F_x = -2 \int_{-\pi/2}^{+\pi/2} \frac{2Mh}{\mu_w} v_x \cos^2 \theta R \, d\theta$$

or

$$F_x = -\pi R(2Mh)v_x/\mu_w \tag{4.4}$$

in the x direction. We have already calculated the driving force on a circular domain in a simple bias field gradient, dB_z/dx, when we considered the measurement of coercive force in § 3.3.5 of Chapter 3. The result was given there as equation (3.17) so that, if we equate this to equation (4.4) and solve

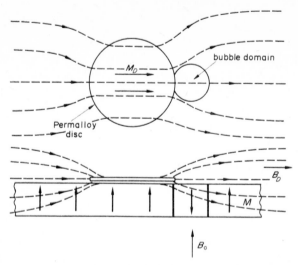

Figure 4.1. The experiment of Rossol and Thiele (1970) in which a bubble domain is made to rotate around the periphery of a permalloy disc by means of a rotating in-plane field, B_D.

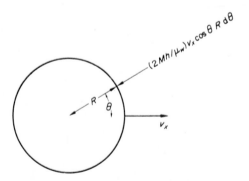

Figure 4.2. A bubble domain moving with a translational velocity v_x experiences a viscous drag which is proportional to the normal component of the wall velocity and acts normally upon the moving wall.

for v_x, we obtain

$$v_x = \mu_w \left(R \frac{\mathrm{d}B_z}{\mathrm{d}x} - \frac{4}{\pi} B_c \right) \tag{4.5}$$

where B_c is the coercive force of the material.

Referring to figure 3.8, we see that $2R(\mathrm{d}B_z/\mathrm{d}x) = \Delta B_z$ is the change in the bias field across the diameter of the bubble domain. Equation (4.5) may thus be written

$$v_x = \frac{\mu_w}{2} \left(\Delta B_z - \frac{8}{\pi} B_c \right) \tag{4.6}$$

which was the form given by Thiele (1971). This result only applies when the bubble domain maintains its circular shape and constant diameter whilst in motion, a situation which applied fairly well in the experiment described by Rossol and Thiele (1970). Their measurements of lag angle as a function of frequency, mentioned above, were used to calculate the wall mobility in $Sm_{0.55}Tb_{0.45}$ and were in excellent agreement with previous measurements of μ_w for straight walls in this material.

Another experiment which involved bubble translation in orthoferrites was done by Copeland and Spiwak (1971) who used the field gradient produced by a current carrying conductor, laid across the diameter of an isolated bubble domain. A sinusoidal current caused the bubble to oscillate along a line at right angles to the conductor and the distance the bubble moved could be measured by observing the bubble domain with the Faraday effect. Velocities up to 80 m/s were observed in $YFeO_3$ and the values of μ_w, again inferred from equation (4.6), were in excellent agreement with previous measurements by Shumate (1971).

4.2.3 *The bubble collapse method of measuring mobility*

The satisfactory agreement between a direct measurement of wall mobility and the inference of this same mobility from a bubble translation experiment suggested that wall mobility was of the utmost importance in the choice of a material for a bubble domain device. This led to a great deal of work being done on the problem of mobility measurement and the technique adopted by the majority of workers in this field was the bubble collapse method described by Bobeck *et al.* (1970). The reason for choosing this method, as opposed to the translational method described in § 4.2.2, was that no permalloy disc or conductor was needed, these being particularly difficult to make to the very small dimensions of the bubble domains in the garnet materials.

In the bubble collapse method of measuring mobility, a bubble domain is set up in a bias field, B_0, which gives it a stable diameter, D_0, somewhere in between the run-out and collapse diameters which were discussed in Chapter 2, § 2.3.6. A pulsed magnetic field, B_P, is then applied, in the same direction as B_0 and of sufficient magnitude and duration to produce bubble collapse. Measurements are made of the amplitudes and durations of the pulsed fields which will just produce collapse and the wall mobility is deduced from these measurements in the following way.

Referring back to figure 2.18, it is clear that if we are driving the bubble domain towards collapse, and assume that the static field solution applies under dynamic conditions, the wall of the bubble domain will be situated in a field equal to the value of the pulsed field, B_P, at the beginning of its motion. At the end of its motion, just before collapse, the wall will be in a weaker field, $B_P - (B_{COL} - B_0)$, having travelled up the curve shown in figure 2.18.

The portion of the curve shown in figure 2.18 which connects collapse and run-out is very nearly a straight line, and figure 2.15 shows that this is a fair approximation for a wide range of values of λ/h. It follows that we may assume a linear variation in the field seen by the wall during the bubble collapse process and that the velocity of the wall should be,

$$v_r = \mu_w \left[B_P - \frac{(B_{COL} - B_0)(D_0 - D)}{(D_0 - D_{COL})} \right] \tag{4.7}$$

from equation (4.2). We are taking the applied fields to be negative here, as in Chapter 2, so that v_r is negative.

Because $v_r = \frac{1}{2}(dD/dt)$, equation (4.7) is a simple first order linear differential equation and gives the diameter as a function of time

$$D = A(e^{-t/T} - 1) + D_0 \tag{4.8}$$

where the characteristic time constant of the process is

$$T = \frac{(D_0 - D_{COL})}{2\mu_w |B_{COL} - B_0|} \tag{4.9}$$

and A is a constant

$$A = \frac{B_P(D_0 - D_{COL})}{(B_{COL} - B_0)} \tag{4.10}$$

The experimental results given by Bobeck *et al.* (1970) are for the collapse time which is the time, $t = T_c$, required for D to fall from D_0 to D_{COL}. Equation (4.8) thus gives

$$D_{COL} = A(e^{-T_c} - 1) + D_0 \tag{4.11}$$

so that, substituting equation (4.10) into equation (4.11) and rearranging gives

$$T/T_c = -1/\log_e \left[1 - \frac{(B_{COL} - B_0)}{B_P} \right] \tag{4.12}$$

Equation (4.12) has been plotted in figure 4.3 with $y = T/T_c$ and $x = B_P/(B_{COL} - B_0)$, the experimental quantities used by Bobeck *et al.* (1970). As $\log_e (1 - 1/x)$ tends to $-1/x$, as x becomes large compared to unity, we would expect equation (4.12) to follow the straight line $y = x$ for large values of x. Figure 4.3 shows that this happens very slowly and, in fact, the slope of equation (4.12) is very close to unity once $x > 2$ and the function is, to all intents and purposes, a straight line. This interesting feature of the solution to the bubble collapse problem was pointed out by Callen and Josephs (1971) in their detailed analysis of the problem which included the effects of a finite coercive force.

It follows that if we make measurements of T_c for various values of B_P, always using the same initial conditions of D_0 and B_0 so that T is a constant, we would expect our results to lie on a curve of the kind shown in figure 4.3.

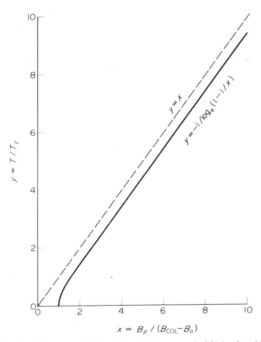

Figure 4.3. The function $y = -1/\log_e(1 - 1/x)$ is remarkable in that it runs almost exactly parallel to $y = x$ once x exceeds about 2. Note that the function leaves the x axis with infinite slope.

For large values of B_P, the results should lie on a straight line and the slope of this will give a measure of μ_w because

$$\frac{d(1/T_c)}{d(B_P)} = \frac{1}{T|B_{COL} - B_0|} \frac{d(T/T_c)}{d[B_P/(B_{COL} - B_0)]} \qquad (4.13)$$

and as $d(T/T_c)/d[B_P/(B_{COL} - B_0)] \approx 1$ we have, using equation (4.9)

$$\mu_w = \tfrac{1}{2}(D_0 - D_{COL})\frac{d(1/T_c)}{d(B_P)} \qquad (4.14)$$

where we have substituted for T from equation (4.9).

The results given by Bobeck *et al.* (1970) for bubble collapse in the ortho-ferrites and in gadolinium terbium iron garnet showed the expected linear relationship between $1/T_c$ and B_P at high values of pulsed field. The slope of the characteristic gave a value of μ_w which was in good agreement with other measurements. At low values of B_P, however, Bobeck *et al.* (1970)

appeared to expect the straight line relationship to continue and cut the x axis at $B_P/(B_{COL} - B_0) = 1$, whereas figure 4.3 shows that the straight line should extrapolate back to $x \approx 0.5$. This would be correct for a negligible coercive force, which is the situation in most bubble domain materials. Callen and Josephs (1971) gave some bubble collapse data for $DyFeO_3$ which showed the correct curvature at low values of pulsed field. The characteristic must leave the x axis with infinite slope, as shown in figure 4.3, because a vanishingly small increase in B_P above $(B_{COL} - B_0)$ will reduce T_c from infinity to a finite value. The value of μ_w given by Callen and Josephs (1971) for $DyFeO_3$ was 3.7×10^4 m/s per T, which agreed reasonably well with the value of 3.3×10^4 measured by Seitchik *et al.* (1971), who used a very elegant method for straight domain walls on the same sample of material.

4.2.4 *The observation of a saturation velocity in bubble domain collapse*

Not all bubble collapse data shows good agreement with the simple wall mobility concept. The results given by Bobeck *et al.* (1970) for the hexagonal ferrites showed that the bubble collapse time fell to a minimum value as the pulsed field was increased, suggesting a limiting or saturation velocity for the domain wall during this kind of dynamic collapse process. Argyle *et al.* (1971) confirmed that there was a well defined saturation velocity for bubble domain collapse in a number of epitaxial garnets. Callen *et al.* (1972) measured a saturation velocity of 12 m/s in $Eu_{0.6}Y_{2.4}Fe_{3.8}Ga_{1.2}O_{12}$ during bubble collapse and a saturation velocity between 8 m/s and 9 m/s was measured by O'Dell (1973) in $Eu_{0.6}Y_{2.4}Ga_{1.1}Fe_{3.9}O_{12}$, using a different technique which involved both radial expansion and radial contraction of the bubble domain.

The problem of the saturation velocity was taken up theoretically by Slonczewski (1972a, b, c) who showed that a limiting velocity would be observed for domain wall motion in bubble domain materials because the wall would become unstable and lose its planar form. The saturation velocities predicted (Slonczewski 1971) were of the same order of magnitude as those observed experimentally (Argyle *et al.* 1971) and suggested that there would be very serious speed limitations on bubble domain devices using these, otherwise excellent, garnet materials. Fortunately, it turned out that the velocity saturation observed for radial contraction or expansion did not apply for a purely translational bubble motion.

4.2.5 *Bubble translation at velocities above the collapse saturation velocity*

The saturation velocities observed in the bubble collapse experiments were of the order of a few metres per second and, if the same limiting velocity were to apply to bubble translation, this would be a very serious limitation upon the speed at which a bubble domain device might operate. There is, of course, no reason to expect that the saturation collapse velocity is going to be the limiting velocity for bubble translation. We shall see in § 4.3.2 that it is not only the value of the local field at the domain wall which is of importance in bubble dynamics, but also the nature of the field gradient. In the majority of bubble translation experiments this field gradient acts in such a way as to inhibit the instabilities described by Slonczewski (1971) and allow much higher velocities to be attained.

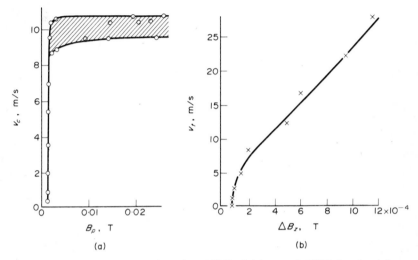

Figure 4.4. The experimental results of Vella-Coleiro *et al.* (1973) (reprinted by permission from *App. Phys. Lett.*). The limiting velocity observed during bubble collapse in a pulsed field, B_p, as shown in (a), does not apply to bubble translation when a field difference, ΔB_z, is established across the bubble diameter, as shown in (b).

Figure 4.4 shows some results by Vella-Coleiro *et al.* (1973) which demonstrate the considerable difference between the velocities which can be obtained in the collapse and translational modes. Figure 4.4(a) shows the bubble collapse results for an epitaxial layer of $Y_{2.4}Eu_{0.6}Ga_{1.2}Fe_{3.8}O_{12}$ and are very similar to the results published by Callen *et al.* (1972), for the same material, in showing a saturation velocity around 10 m/s. Figure 4.4(b) shows the bubble translation results, obtained by means of the permalloy disc technique of Rossol and Thiele (1970) which was described here in § 4.2.2. The translational velocity may be seen to reach a value well above 10 m/s at the higher drive fields and shows no sign of running into saturation, although a higher bubble mobility does seem to apply for lower drive fields. These are very encouraging results for the bubble domain device. A velocity of 20 m/s for a 5 μm diameter bubble domain means that the bubble may propagate through four diameters, which would be a typical periodic length for a propagation track, in a time of 1 μs. This is a data rate of 1 MHz which is the order of magnitude desired.

Bubble translation experiments on other materials have shown very similar results to those shown in figure 4.4(b). Vella-Coleiro and Tabor (1972) measured translational velocities up to 5 m/s in an epitaxial layer of $EuEr_2Ga_{0.7}Fe_{4.3}O_{12}$ using the field gradient produced by a small set of current carrying conductors. Bonner *et al.* (1972) made measurements on a series of yttrium europium aluminium garnets which had small quantities of ytterbium substituted into them. Velocities up to 16 m/s were observed, these results being of particular interest because the ytterbium was found to linearise the velocity against ΔB_z characteristic and increase the bubble mobility at high drive fields.

4.2.6 *Experiments involving propagation tracks*

The experiments described in the previous section were directed towards measuring the translational mobility of the bubble domain in a situation similar to that found in a device where the bubble is moving along contained by a bias field minimum moving at the same velocity. An obvious extension of this idea is to measure the dynamic properties of a material by actually using it in some kind of bubble domain device and this approach has been used by a number of authors.

Stroboscopic techniques were used by Rossol (1971), Moore (1971) and by Yoshimi *et al.* (1972) to observe the continuous motion of bubble domains along magnetic overlay and current carrying conductor propagation tracks. Fischer (1971) used a magnetic overlay track but, instead of using a continuously rotating field, a pulsed in-plane field was used so that a bubble domain could be taken over just two periods of the propagation track and an observation of its position could be made before and after the pulse program to see if correct transfer had been made.

Experiments of this kind really test both the propagation track and the material so that we shall postpone any further discussion until Chapter 5 where the magnetic overlays will be considered in detail.

4.2.7 *Experiments involving the dynamics of 'hard' bubbles*

The exact nature of the wall structure of a bubble domain was discussed briefly in Chapter 2, § 2.3.3, where figures 2.13 and 2.14 were used to illustrate the possibility of a segmented wall structure as opposed to a simple Bloch wall surrounding the bubble.

The possibility of a complex wall structure was first proposed by Tabor *et al.* (1972) and by Malozemoff (1972) in order to explain the fact that bubble domains were observed, in some materials, to collapse over a range of applied bias fields, instead of all collapsing at exactly the same B_{COL} which would be expected. Bubble domains which were observed to persist at bias fields well above the expected collapse field were termed 'hard' bubbles and their behaviour was explained by the complex wall structure.

This is illustrated in figure 4.5 where we have repeated the proposed structure of figure 2.14 to show that, as the bubble domain contracts, a larger proportion of its wall volume is occupied by the regions in which the magnetisation rotates about the θ axis, as opposed to rotating about the

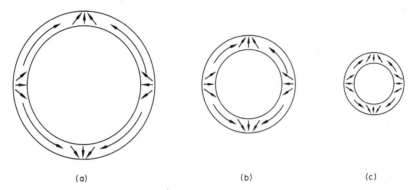

Figure 4.5. As a hard bubble contracts, in an increasing bias field, the total energy may increase even though the wall length decreases.

r axis as it does in the simple Bloch wall. These regions are termed Bloch lines and have a higher energy density than the simple Bloch wall regions because the atomic magnetic moments must turn through a greater angle as we go from lattice point to lattice point. It follows that the total wall energy is no longer falling as the bubble radius falls but may even increase as the Bloch lines are forced close to one another.

The static properties of the hard bubble domain present no difficulties from the device point of view because, if the model of a segmented wall is the explanation of the high collapse field, a device would normally operate at a bubble diameter where the volume of the wall was predominantly occupied by a simple Bloch wall and the fact that the bubbles passing through the device might have different numbers of segments would produce a negligible difference in their behaviour. It might even be argued that a degree of 'hardness' is desirable in order to extend the operating margins of the device.

The dynamic properties of hard bubbles present a very different picture but it is at this point we must remember exactly what the experimental facts are. The hard bubble is one which collapses at an unexpectedly high bias field. It is observed more readily in certain materials and usually has to be generated in a very particular way. Its static and dynamic properties have been attributed to a complex wall structure but, so far, the wall structure has never been observed in any of the bubble domains which were used in the dynamic experiments we shall discuss. It follows from this that, while the properties of the hard bubble may be very interesting, there is no reason to expect bubbles with these unusual properties to occur in bubble domain devices.

The most interesting and detailed study of the static and dynamic properties of hard bubbles was published by Slonczewski and Malozemoff (1972). These authors found that hard bubbles could be generated in epitaxial layers of europium yttrium gallium iron garnet, which contained small amounts of terbium or ytterbium, by means of the pulsed field technique which was described here in connection with figure 1.7. Hard bubbles were selected by simply increasing the bias field well above the expected collapse field and working with the bubble domains which remained.

The most striking feature of dynamic measurements made upon these hard bubbles by Slonczewski and Malozemoff was found when they attempted to make bubble collapse measurements of the kind described in §§ 4.2.3 and 4.2.4. When very large pulsed fields were applied, both the normal bubbles and the hard bubbles showed a collapse time which did not depend very much upon the applied field but the hard bubble took as long as ten times the time to collapse. The results taken at pulse fields just sufficient to produce collapse also showed a dramatic difference; the hard bubble appeared to have a much lower value of μ_w, as given by equation (4.14).

The dynamic behaviour of hard bubble domains in a bias field gradient was also found to show some very interesting features. These had been reported by Tabor *et al.* (1972), Vella-Coleiro *et al.* (1972), Malozemoff and Slonczewski (1972) and Slonczewski and Malozemoff (1972). The main effect which was discussed by these authors was that hard bubbles appeared to move at an angle to the direction of the applied bias field gradient. It was pointed out by Slonczewski (1972d) and Slonczewski and Malozemoff (1972) that we should not expect the so-called normal bubble, the one shown in figure 2.13, to move exactly parallel to the direction of the field gradient because it is not a symmetric structure when we consider a direction of motion passing through its diameter. By contrast, the hard bubble can be symmetric and this is illustrated in figure 4.6. Some experimental results given by Slonczewski and Malozemoff (1972) support this idea in that these authors measured the angle at which bubble domains of varying degrees of hardness moved relative to the direction of the gradient and were able to deduce that there was some evidence that the bubble domain with two Bloch lines, as shown in figure 4.6(b), moved parallel to the field gradient.

In the bubble domain device, the bubble moves in a moving field gradient and its direction is controlled by gradients which are perpendicular to the direction of motion. For this reason, it is not very important that the bubble domain may move at an angle to a simple one dimensional field gradient because such a gradient would never be found in a practical device. The low mobility of hard bubbles in translation, however, is of great importance from the device point of view (Tabor *et al.* 1972, Vella-Coleiro *et al.* 1972 and Slonczewski 1972d) and a considerable amount of work has been done

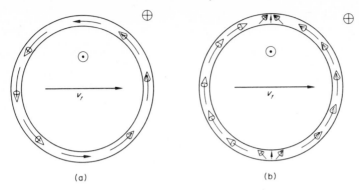

Figure 4.6. In translation, the normal bubble domain, (a), presents an asymmetric structure and the atomic magnetic moments within the wall cannot rotate in unison, as they do when the bubble expands or contracts. In contrast, a hard bubble, (b), is symmetric about the direction of motion.

on finding ways of suppressing the occurrence of hard bubbles by modifications to the material structure (Bobeck *et al.* 1972, Wolfe and North 1972 and Rosencwaig 1972). These techniques are directed towards ensuring that the bubble with only two Bloch lines, as shown in figure 4.6(b), is the preferred variety. It should always be borne in mind, however, that there is no reason to expect hard bubbles to be generated in a bubble domain device because the device does not normally use very intense pulsed fields in its generator.

4.2.8 *Conclusions to be drawn from the dynamic experiments*

When we look back at the dynamic experiments which have been described above, it is clear that all the dynamic data on bubble domains have been obtained by means of three distinct experimental techniques. These are the bubble collapse technique, bubble translation in a bias field gradient and bubble translation by means of a moving field well. We shall now look at these three experimental situations critically and pay particular attention to the nature of the magnetic field gradient seen by the wall of the bubble domain, because it is this field gradient which may explain some of the apparent differences in the results which have been obtained by these three methods.

In the bubble collapse experiment, an isolated bubble domain is held in stable equilibrium with a bias field, B_0, and then this bias field is suddenly increased by means of a pulsed field, B_p. The situation is illustrated in figure 4.7(a). Initially, the entire wall of the bubble domain finds itself in a local field B_p and the bubble begins to contract at a velocity $\mu_w B_p$. As shown in our analysis of the problem in § 4.2.3, the local wall field gets smaller as the bubble contracts. This means that, to a first approximation, the wall is moving in a local field which gets smaller along the direction of motion and

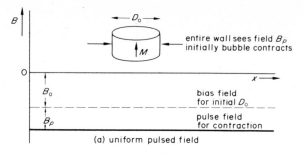

(a) uniform pulsed field

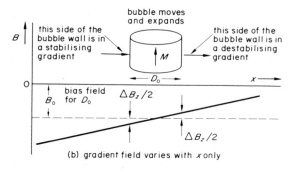

(b) gradient field varies with x only

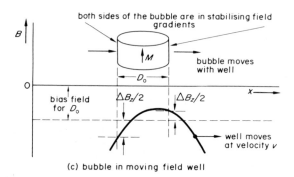

(c) bubble in moving field well

Figure 4.7. Three kinds of bubble domain motion. Attention is directed here towards the field gradient seen by the moving wall.

its motion should, therefore, be stable. If the wall moves too fast, it will see a smaller field and slow down. If the wall moves too slowly, it will see a larger field and speed up.

The bubble translation experiment which uses a simple field gradient, as shown in figure 4.7(b) is a completely different situation compared to bubble collapse when we consider the local field gradient at the wall. Figure 4.7(b) shows the initial situation in such an experiment where a bubble domain has been set up in a bias field B_0 and then a gradient field is established which adds to B_0 on one side of the bubble domain and subtracts from B_0 on the other side. The change in B_z across the bubble diameter is shown as ΔB_z in figure 4.7(b) to correspond to our analysis of this problem in § 4.2.2, which resulted in equation (4.6), where it was assumed that the bubble domain remained stable and circular so that no discussion of the local field gradient at the wall was given. When we look at this local field gradient in figure 4.7(b) it is clear that the left hand side of the bubble domain is, initially, in the same kind of field gradient as we had in the bubble collapse situation. It is moving into a weaker field. The right hand side of the bubble, the leading side, is in just the opposite situation because it is moving into a stronger field and would tend to accelerate. Qualitatively, it is obvious that a bubble domain in the situation shown in figure 4.7(b) cannot remain circular and it would seem very likely that its wall would become unstable on the leading side. We shall see that this can be shown quantitatively in § 4.3.2.

Finally, we consider the third kind of experiment which is illustrated in figure 4.7(c) and is the case of bubble translation in a moving field well. We again assume a bias field B_0 exists for the initial diameter D_0 and, initially, we superimpose a parabolic field well, which should be imagined as the surface of revolution of the curve shown in figure 4.7(c), in such a way that the local field at the wall remains at B_0 while the total bias field is reduced inside the bubble and increased outside. Bubble motion is then produced by moving the entire field well along at velocity v and figure 4.7(c) supposes that this would require that the bubble domain lags behind the field well by a constant distance so that the same ΔB_z is generated across the bubble diameter as we have in the case of figure 4.7(b). The exact situation is more complicated, and will be considered in § 4.3.3, but we can see at once that both sides of the bubble domain are now moving in the correct kind of local field gradient which is needed for stable motion. That is, the local field decreases in the direction of motion, this local field being the difference between the well field and the wall field of the bubble which is equal and opposite to the value of B_0 required for the bubble diameter D_0, as shown in figure 4.7(c). If either side of the bubble domain moves at the wrong velocity, there will be a stabilizing effect in that the field will change in such a way as to restore the correct velocity.

The conclusion to be drawn here is that we shall have to look very closely at the stability of motion for bubble domains in field gradients because we can only assume that the bubble remains circular when the conditions for stability apply. If the bubble becomes distorted in motion, it will be necessary to calculate the change in the wall field produced by this distortion and so find the change in the field responsible for the motion. If the bubble domain becomes completely unstable, however, there is no point in considering its motion and the problem becomes one of finding the correct kind of stabilising field for the motion which is required.

4.3   Bubble Domain Motion in a Moving Field Gradient

4.3.1   *Introduction*

We shall now attempt to build up a theoretical model for bubble domain
motion which is applicable to bubble domain devices and the problems of
the magnetic overlay patterns, used as propagation tracks in these devices,
which we shall consider in Chapter 5. This means that we shall have to
concentrate upon the stability of the domain in motion and its changing
shape, while adopting a very simple model for the fundamental atomic
scale processes that are taking place within the moving wall of the domain
itself. From a device point of view, this is a satisfactory standpoint because
we know experimentally that conditions can be set up which cause stable
translation of a bubble domain at the velocities which are needed. The
problem is to understand what defines these conditions.

Bubble translation experiments which give a reasonably linear and well
defined relationship between bubble domain velocity and the velocity of a
moving field gradient were reviewed above in § 4.2.5. These were the results
given by Vella-Coleiro *et al.* (1973) for bubble motion in a garnet layer by
means of the moving field well around a permalloy disc, reproduced here
as figure 4.4, and the results given by Vella-Coleiro and Tabor (1972) for
translation in the field gradient produced by a transmission line. The latter
experiment was particularly elegant in that the local bias field was increased
as the bubble domain moved so that a true moving gradient field was produced, of the kind shown in figure 4.7(b) with the gradient itself being trans-
lated to the right with the bubble domain. This maintains the bubble at
constant diameter, to a first approximation.

The first problem to be considered here is the effect of the field gradient
upon the stability of the bubble domain itself. This will be introduced by
considering the static stability problem and then we shall look at bubble
domain motion in a moving field well which should ensure complete stability. A very interesting inertial effect will be found: the moving bubble
domain has an effective mass, which is due to the fact that the bubble must
contract as it moves and that this contraction depends upon the square of
the velocity. A computer simulation of bubble domain translation then lays
the foundation we need for Chapter 5 where the dynamic problems associated
with magnetic overlays will be considered.

4.3.2   *Bubble domain stability when there is a field gradient at the wall*

Bubble stability was discussed qualitatively in § 4.2.8 from the point of view
of bubble motion in a field gradient or field well. As a first step in understanding this problem quantitatively, it is as well to look at static stability
in a field gradient. This also leads to a result that will be needed for the discussion of magnetic overlays given in Chapter 5.

We first consider a situation, similar to that shown in figure 4.7(c), in which the bubble domain is situated in a field well and both the well and the bubble are at rest. If the field well is of such a shape that the magnitude of the field at the wall is B_0, the uniform bias field required for the domain diameter D_0, then the bubble domain will be in equilibrium with the field. Our problem is to calculate the collapse and run-out diameters which apply in this situation and see how and why they differ from the collapse and run-out diameters calculated in Chapter 2, § 2.3.6.

The new collapse diameter is easily found by taking our previous result, equation (2.39), for the effective field at the wall due to the magnetisation of the bubble and its wall energy. This may be equated to the bias field, as before, which is now a function of the coordinate r. Thus

$$\frac{1}{1 + q(D/h)} - \frac{(\lambda/h)}{(D/h)} = \frac{-B(r/h)}{\mu_0 M} \qquad (4.15)$$

is now the equilibrium condition and, as figure 4.8 shows, bubble collapse will occur when the derivative of the left hand side of equation (4.15), with respect to $r/h = D/2h$, is equal to the derivative of the right hand side. The broken line in figure 4.8 is $-B(r/h)/\mu_0 M$, to correspond to our previous diagrams in Chapter 2 where $-B_0$ was shown, and is thus referred to as a

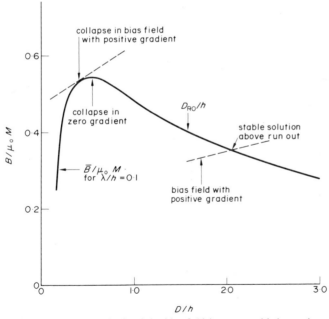

Figure 4.8. When the magnitude of the bias field increases with increasing radius, the collapse diameter of the bubble domain is smaller and the run-out diameter is larger than in a uniform bias field.

bias field with a positive gradient for convenience. As shown in figure 4.7(c), the radial gradient is negative because the bias field is negative.

It is clear from figure 4.8 that bubble collapse must now occur at a smaller bubble diameter than in a uniform bias field. The opposite applies when $-B(r/h)$ has a negative gradient and this situation is shown in figure 4.9.

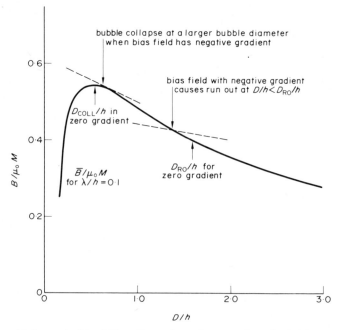

Figure 4.9. A negative bias field gradient reduces the range of possible bubble domain diameters.

Actual numerical values of D_{MIN}/h are easily calculated numerically from our previous evaluations of the left hand side of equation (4.15), which were shown in figure 2.15 for a number of values of λ/h. Two examples are shown in figures 4.10 and 4.11 for $\lambda/h = 0.1$ and $\lambda/h = 0.3$ where it can be seen that the positive gradient has very little effect upon the collapse diameter but the negative gradient may have a very dramatic effect upon the collapse diameter as λ/h becomes large.

The very existence of a negative gradient requires some comment because we are concerned at the moment with a symmetrical field well and when this takes on what we are calling a negative gradient the situation is, in fact, a field of the kind shown in figure 4.12. As shown in this figure, motion to the left will cause the left hand side of the bubble wall to see a reduced bias field, and thus expand or move to the left, while the right hand wall will see an increased bias field and contract, again moving to the left. The bubble

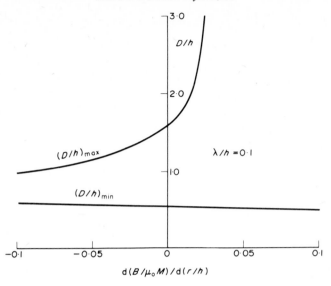

Figure 4.10. The collapse, $(D/h)_{min}$, and run-out, $(D/h)_{max}$, diameters as a function of bias field gradient for $\lambda/h = 0 \cdot 1$.

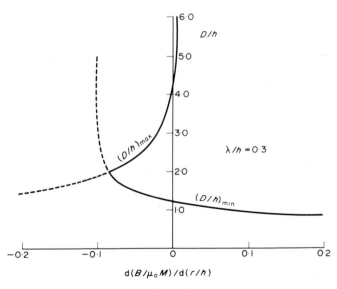

Figure 4.11. As figure 4.10 but for $\lambda/h = 0 \cdot 3$. Quite a modest negative bias field gradient makes the very existence of a bubble domain impossible.

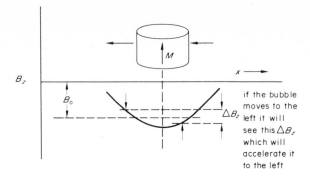

Figure 4.12. Positional instability when the field gradient is negative at all points around the bubble domain wall.

domain is thus completely unstable from a positional point of view. The relationships shown in figures 4.10 and 4.11 would thus only apply if the position of the bubble domain were to be maintained by some other stabilising force and this would be the situation in a linear gradient of the kind shown in figure 4.7(b). In that case the left hand side of the bubble domain is stabilised by means of a positive field gradient and the instability of the right hand side would be caused by the negative field gradient, with respect to r, and follow the relationships shown in figures 4.10 and 4.11.

We may now turn our attention towards the run-out of the bubble domain when there is a field gradient at the wall. First, consider the situation in which the bubble domain is contained within a field well, as shown in figure 4.7(c). If the bubble tries to expand, it sees an increased bias field and we would, therefore, expect run-out to occur at a greater diameter than in the case of a uniform bias field because this increase in the bias field will produce an additional restoring force that will maintain the original diameter of the bubble domain.

From this point of view, it is necessary to return to Chapter 2 where run-out was analysed in § 2.3.5. In that paragraph we dealt with the problem of run-out by asking what field would be required in order to produce pull-in for an elliptical domain which co-existed with a bubble domain just supported by the run-out field. We now look at this problem the other way round and ask why it is that a bubble domain, which is circular, does, in fact, run out when the bias field is sufficiently reduced. The answer to this question is that the wall energy was seen to be responsible, in Chapter 2, for an effective wall field

$$\frac{B_w}{\mu_0 M} = \frac{-\sigma_w}{\mu_0 M^2 h}(D/h)^{-1} \tag{4.16}$$

and this may be interpreted as a field which depends upon the radius of curvature, $\rho \to D/2$, of the bubble domain wall, so that

$$\frac{B_w}{\mu_0 M} = \frac{\sigma_w}{\mu_0 M^2 h}(2\rho/h)^{-1} \qquad (4.17)$$

illustrates the point to be made. This field, B_w, is negative, or in the same direction as the bias field which produces contraction. It follows that an increase in the radius of curvature produces a reduction in B_w, or a field which tends to expand the bubble domain, while a reduction in ρ produces a field which tends to increase ρ. The field, B_w, thus has a stabilising field gradient which is equal to

$$d(B_w/\mu_0 M)/d(r/h) = 2(\lambda/h)/(D/h)^2 \qquad (4.18)$$

from equation (4.16), where we have considered $r \equiv \rho \equiv D/2$ and substituted equation (2.38) so that the material under consideration is defined by the characteristic length λ.

Our argument proceeds by saying that, if we now substitute values of the run-out diameter, D_{RO}/h, for various values of λ/h, into equation (4.18), these values being shown in figure 2.21, we shall obtain a set of field gradients which must be equal to the destabilising gradient produced by the bubble domain magnetisation itself. An example is shown in figure 4.13 for the

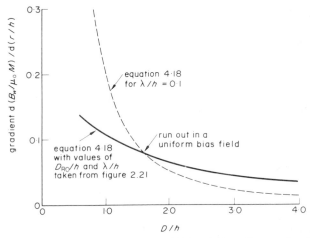

Figure 4.13. The broken curve shows the stabilising field gradient produced by the wall energy for a bubble domain when $\lambda/h = 0.1$. The full curve is the destabilising gradient of the field due to the bubble magnetisation. In the absence of any other field gradient, run-out occurs when these two are equal.

particular case of $\lambda/h = 0.1$. The full line shows the destabilising gradient produced by the bubble domain magnetisation, the broken line shows the stabilising gradient produced by the wall energy. In a uniform bias field, these two gradients are equal and opposite to one another at the run-out diameter.

It is clear from figure 4.13 that, when a bubble domain is held by a bias field at a diameter below run-out, run-out may be precipitated by the application of a negative field gradient at the wall which is equal to the difference between the two curves, shown in figure 4.13, at that particular diameter. By calculating this difference at various values of D/h, we obtain the curve, $(D/h)_{MAX}$, shown in figure 4.10.

Figure 4.13 also shows that a bubble domain may be held in stable equilibrium well above the normal run-out diameter by applying a positive field gradient at the wall. The value of this gradient is the difference between the two curves shown in figure 4.13 again and this enables us to give a complete set of results for $(D/h)_{MAX}$, shown in figure 4.10, covering both positive and negative wall field gradients for the particular case of $\lambda/h = 0.1$.

Figure 4.11 shows the results of similar numerical calculations for the case of $\lambda/h = 0.3$. The range of possible bubble diameters is now seen to be radically reduced when the gradient of the field at the wall is negative and it is not possible to have any bubble domains at all when the gradient is more negative than about 0.08. This is a dimensionless gradient and, would imply a true field gradient of about 3.5×10^{-4} T per μm for the Eu_1Er_2 $Ga_{0.7}Fe_{4.3}O_{12}$ garnet layer used by Vella-Coleiro and Tabor (1972) in their experiments on bubble domain translation in which a negative, or destabilising, field gradient did exist at one side of the bubble. The maximum field gradient used in these experiments was 2.4×10^{-4} T per μm and the value of λ/h is 0.175 when the data given by Shumate *et al.* (1973) is taken into account. From this we may conclude that the experimental situation described by Vella-Coleiro and Tabor (1972) was well within the range of good stability. The same applies to the experiments of Slonczewski and Malozemoff (1972), that were discussed in § 4.2.7 and illustrated the anomalous behaviour of hard bubble domains. The field gradients used and the value of λ/h were both small enough to allow a good range of $(D/h)_{MIN}$ to $(D/h)_{MAX}$.

The field gradients which may occur in the magnetic overlay propagation tracks, considered in Chapter 5, are very much larger than those discussed above and the domain stability problem will be found to be of the utmost importance. Figures 4.10 and 4.11 make it clear that it is only the negative field gradient which is going to be of concern. The positive field gradient increases the stable range of possible bubble diameters.

4.3.3   *Bubble domain motion in a moving field well*

It has been shown in the previous section that bubble domain stability is assured provided that there is a positive field gradient, in the sense of figure 4.7(c), at the domain wall. This leads us to expect that a bubble domain will move along with a moving field well in a stable manner and this problem will now be considered in some detail.

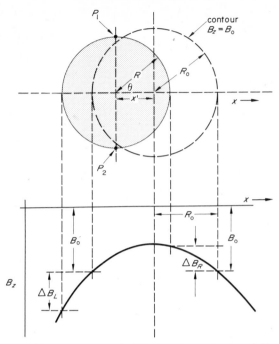

Figure 4.14. Looking down upon a bubble domain moving in a field well which is itself moving, we see the bubble lags behind the contour $B_z = B_0$ which it occupied when at rest.

Figure 4.14 shows the essential features of this dynamic situation. When it is at rest, the bubble domain has a radius R_0 and the domain wall will be situated in a local bias field B_0 so that it is in equilibrium. This is the equilibrium condition that was defined by equation (2.39) and may be written as

$$\frac{1}{1 + q(2R_0/h)} - \frac{\lambda/h}{(2R_0/h)} = \frac{-B_0}{\mu_0 M} \qquad (4.19)$$

where $q = 0.726$.

When the bubble domain is in motion, it will lag behind the moving field well by a distance x', as shown in figure 4.14, so that the wall sees an accelerating local field. This is shown in figure 4.14 as ΔB_L on the left and ΔB_R on the right, simply for the diameter which lies along the direction of motion. These two local fields will only tend to be equal for $x \ll R_0$ and, even then, the local field at other points around the wall may not be compatible with it maintaining its circular shape while in motion. This is the first point which must be brought into the analysis. Figure 4.14 also illustrates a second point. This is that the moving bubble domain must become smaller. This is because the points on the bubble domain wall, labelled P_1 and P_2 in figure 4.14, which have no normal component of velocity, must be in equilibrium with the local field at these points. If the radius of the domain remained at R_0, the field at points P_1 and P_2 would be higher than B_0 and the domain would contract. Figure 4.14 shows P_1 and P_2 at a radius, with respect to the centre of the field, only just greater than R_0 because this would be the situation when the field gradient at the wall is high.

The reduction in the bubble domain radius as a result of its motion is of the greatest importance because this means that its magnetic energy has increased. This effect may be expressed quite simply for small changes in the bubble radius because we know that the total force on the domain wall, due to the magnetisation of the domain and the wall energy is

$$F = (2Mh)(-B_0)(2\pi R_0) \qquad (4.20)$$

from equations (2.34) and (2.13). This force must be equal to $-\partial E/\partial R$, where E is the total energy, so that a small change in domain radius, ΔR, will produce a change in domain energy of

$$-\Delta E = (2Mh)(-B_0)(2\pi R_0)\,\Delta R \qquad (4.21)$$

If we now consider the case of a field well which has a very high gradient at the domain wall, so that the points P_1 and P_2 would be at a radius very close to R_0, with respect to the field centre, the change in radius, ΔR, will be related to the change in bubble position, x', as

$$-\Delta R = (x')^2/2R_0 \qquad (4.22)$$

The velocity of the domain, if we assume a simple mobility and let $\Delta B_L \approx \Delta B_R \approx (dB_z/dx')x'$, in figure 4.14, is given by equation (4.6), for small x', and will be

$$v_x = \mu_w(dB_z/dx')x' \qquad (4.23)$$

for negligible coercive force. Substituting equations (4.22) and (4.23) into equation (4.21) then gives

$$\Delta E = \frac{\pi(2Mh)(-B_0)v_x^2}{\mu_w^2(dB_z/dx')^2} \qquad (4.24)$$

It follows that the bubble domain energy increases with the square of its velocity and that we may attribute an effective mass to the moving bubble. This means that there will be inertia-like effects in that a finite time will be required to change the bubble domain velocity when it is moving along with a field well of this kind. The time constant which applies in this situation may be estimated when we write down the complete equation of motion as the sum of the inertial force

$$F_i = -\partial(\Delta E)/\partial x \qquad (4.25)$$

from equation (4.24), the viscous drag force, F_x given by equation (4.4), and the local field force, F_B, which will depend upon the exact shape of the field well under consideration. Collecting these results the equation of motion is

$$-\frac{2\pi(2Mh)(-B_0)}{\mu_w^2(dB_z/dx')^2}v_x\frac{dv_x}{dx} - \frac{\pi(2Mh)R}{\mu_w}v_x + F_B = 0 \qquad (4.26)$$

which, as $v_x = dx/dt$, is first order in v_x and has the form

$$\tau\, dv_x/dt + v_x = F_B(t) \qquad (4.27)$$

where $F_B(t)$ is a function of time describing the, as yet, unspecified field well. The time constant τ is given by

$$\tau = \frac{h}{\mu_w\mu_0 M}\left[\frac{2(-B_0/\mu_0 M)}{(R/h)[d(B_z/\mu_0 M)/d(x/h)]^2}\right] \qquad (4.28)$$

4.3.4 *The inertial time constant*

The inertial time constant we are discussing here, τ, is due to the change in the bubble domain radius and is particular to the bubble domain itself and the shape of the field well in which it is situated. There is no connection here with the classical wall inertia of magnetism, first discussed by Döring (1948) and shown by Henry (1971) to be important only at high frequencies in bubble domain devices. Höfelt (1973) also introduced the classical wall inertia into his study of the dynamics of bubble domain lattices and did not consider the inertial effects brought about by the bubbles changing size.

The result given here, as equation (4.28), is only an approximation valid for large field gradients at the domain wall. A detailed calculation of the time constant, τ, will be given in the next section for a specified field well and it will be seen there that the dimensionless term in square brackets in equation (4.28) may take a value $\approx 10^2$ in a practical situation. It is thus essential to estimate the fundamental time constant $\tau^* = h/\mu_w\mu_0 M$, for a typical material before embarking on this detailed analysis. We know, from the work of Slonczewski (1971), which was mentioned in §4.2.4, that the simple concept of a wall mobility does not apply at very high velocities so that if τ^* is found to be very small, compared to a typical bubble domain device operation time, there is no point in looking at this inertial effect in more detail.

Table 4.1 summarises some published data on a number of materials. It may be seen that the value of $h/\mu_w\mu_0 M$, which comes into equation (4.28), is of the order of 10 nS for most typical bubble domain layers so that if the dimensionless part of equation (4.28) can have a value as high as 10^2, we shall be faced with an inertial time constant of about 1 μs. This would mean that inertial effects would be important in practical bubble domain devices and it is essential to look at this problem more carefully.

Table 4.1.

Material	h, μm	$\mu_0 M$, T.	μ_w, m/s per T.	τ^*, ns	Ref.
$Gd_{2.3}Tb_{0.7}Fe_5O_{12}$	12.5	0.0142	1.08×10^4	82	Bobeck et al. (1970)
$Y_{2.4}Eu_{0.6}Ga_{1.1}Fe_{3.9}O_{12}$	10.8	0.0210	$\approx 10^4$	≈ 50	Argyle et al. (1971)
$Y_{2.53}Gd_{0.47}Ga_{1.05}Fe_{3.95}O_{12}$	3.7	0.0150	$> 5 \times 10^4$	< 5	
$TmFeO_3$	60	0.0140	6.3×10^4	68	Seitchik et al. (1971)
$DyFeO_3$	42	0.0128	3.3×10^4	100	
$Gd_{2.34}Tb_{0.66}Fe_5O_{12}$	15.0	0.0168	0.5×10^4	180	Fischer (1971)
$Eu_2ErGa_{0.66}Fe_{4.34}O_{12}$	15.0	0.0244	1.8×10^4	34	
$EuEr_2Ga_{0.7}Fe_{4.35}O_{12}$	4.0	0.0182	1.8×10^4	12	
$Sm_{0.55}Tb_{0.45}FeO_3$	50.0	0.0126	9.0×10^4	44	
$EuEr_2Ga_{0.7}Fe_{4.3}O_{12}$	5.7	0.0250	0.96×10^4	24	Vella-Coleiro and Tabor (1972)
$YGdTm(GaFe)_5O_{12}$	3.9	0.0163	4.1×10^4	5.8	Vella-Coleiro (1972)
$Y_{2.4}Eu_{0.6}Ga_{1.2}Fe_{3.8}O_{12}$	8.1	0.0196	16.4×10^4	2.5	
$YEu_{1.8}Tm_{0.2}(AlFe)_5O_{12}$	11.3	0.0210	4.7×10^4	11.5	
$YEu_{1.8}Yb_{0.15}(AlFe)_5O_{12}$	13.0	0.0205	1.3×10^4	49	
$YEu_{1.99}Ho_{0.01}(AlFe)_5O_{12}$	9.0	0.0169	1.66×10^4	32	
$YEu_{1.99}Pr_{0.01}(AlFe)_5O_{12}$	7.2	0.0254	8.0×10^4	3.5	

4.3.5 *Computer simulation of bubble domain translation in a parabolic field well*

The motion of a bubble domain within a moving field well of the kind shown in figure 4.14 may be analysed numerically once the field well is defined. The simplest assumption is to suppose a parabolic shape for the well; that is

$$\frac{B_z}{\mu_0 M} = \frac{B_0}{\mu_0 M} + \frac{B_1}{\mu_0 M}\left(\frac{r^2 - R_0^2}{h^2}\right) \tag{4.29}$$

where B_0 and B_1 are negative for the situation shown in figure 4.14 and r is the distance of the point of reference from the axis of symmetry of the field well. As a first step in the calculation, we solve equation (4.19) for the radius of the bubble domain when it is at rest, R_0. The bubble is then moved back a distance x' and we use the fact that the points P_1 and P_2, in figure 4.14, must remain in equilibrium with the field so that

$$\frac{1}{1 + q(2R/h)} - \frac{\lambda/h}{(2R/h)} + \frac{B_0}{\mu_0 M} + \frac{B_1}{\mu_0 M}\left(\frac{r^2 - R_0^2}{h^2}\right) = 0 \tag{4.30}$$

where

$$r^2 = R^2 + x'^2 \tag{4.31}$$

must be solved numerically to give the new radius R. Equation (4.30) is obtained from equations (4.19) and (4.29) while equation (4.31) defines the position of the points P_1 and P_2, where the normal component of the wall velocity is zero, relative to the axis of the field well.

Having obtained the new, smaller, bubble domain radius we calculate the value of the applied field at the wall of the domain, which is at a distance

$$r^2 = (R^2 + x'^2 - 2Rx' \cos \theta) \tag{4.32}$$

and subtract this field from the equilibrium field at $r^2 = R^2 + x'^2$ to find the field responsible for the velocity of the wall, $v_n = \mu_w B$. The result is

$$B = (2B_1 Rx' \cos \theta)/h^2 \tag{4.33}$$

which indicates a particular property of the parabolic well in that it supports a circular bubble domain in motion because $v_x = v_n/\cos \theta$ so that

$$v_x = 2\mu_w B_1 Rx'/h^2 \tag{4.34}$$

showing that x' and v_x are very nearly linearly related, because the change in R is small.

The final step in the calculation is to examine the relationship between v_x and the change in bubble domain radius, which we expect to be parabolic in the sense that

$$(R_0 - R)/h = C(v_x/\mu_w\mu_0 M)^2 \tag{4.35}$$

where C is a constant to be determined numerically.

Figure 4.15 shows some results of such a computer simulation of bubble motion in a parabolic well, using parameters to describe the applied field and a material characteristic length, λ, which would be sensible for a practical device situation. It is clear that equations (4.34) and (4.35) do describe the situation quite well.

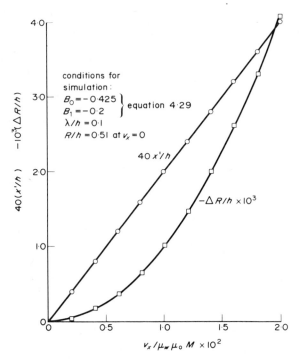

Figure 4.15. Computer simulation of bubble domain translation in a moving parabolic field well. ($\mu_0 M$ has been set equal to unity).

Substitution of equation (4.35), for $\Delta R = (R_0 - R)$, into equation (4.21) thus gives the numerical constant relating ΔE and v^2 for the particular conditions of simulation. The same steps, equations (4.25), (4.26) and (4.27), lead to the time constant

$$\tau = \frac{h}{\mu_w\mu_0 M}[4C(-B_0/\mu_0 M)] \tag{4.36}$$

being defined in terms of the numerically determined constant C and the applied bias field.

The results of the whole series of simulations are shown in figure 4.16, where a value of $\lambda/h = 0.1$ has been chosen so that the bubble diameter to height ratio is near unity; the situation usually required in a bubble domain device. It may be seen that the value of τ/τ^* becomes considerable when the field gradient at the wall of the bubble domain is small; a result which would be expected in view of equation (4.28). The rapid decrease in τ/τ^* with increasing radius would also be expected from equation (4.28) as a term $(R/h)^3$ appears in the denominator when $B_z/\mu_0 M$ is parabolic.

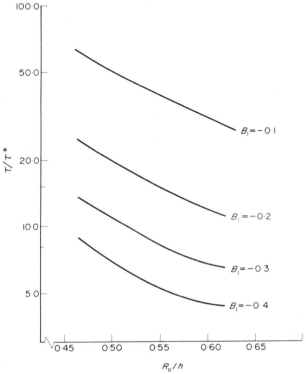

Figure 4.16. Computer simulation of bubble domain translation in a moving field well which is parabolic. The inertial time constant, τ, may be very much greater than $\tau^* = h/\mu_w\mu_0 M$. ($\mu_0 M$ has been set equal to unity).

As an example of the significance of the results shown in figure 4.16, consider the layer of $Eu_{0.6}Y_{2.4}Ga_{1.1}Fe_{3.9}O_{12}$ described in Table 4.1 which is 10·8 μm thick and has $\mu_0 M = 0.021$ Wb/m^2. The value of τ^* for this layer is ≈ 50 nS and, if it were used in a device having bubbles with $D/h = 1$ and a gradient parameter $B_1 = -0.1\mu_0 M$, figure 4.16 shows that the inertial time constant would be ≈ 2.5 μS. This would have a significant effect on the device behaviour if the data rate approached 250 kHz. It should be noted that, when $B_1 = -0.1\,\mu_0 M$, the true field gradient at the wall is $\approx 2 \times 10^{-4}\,\text{Wb/m}^2$ per micron, in this particular material, which is by no means a small gradient as we shall see when magnetic overlays are considered in Chapter 5. Still considering this particular layer of $\text{Eu}_{0.6}\text{Y}_{2.4}\text{Ga}_{1.1}\text{Fe}_{3.9}\text{O}_{12}$, the value of $\mu_w \mu_0 M$ is $\approx 210\,\text{m/S}$, from Table 4.1, so that our computer simulation has taken the velocity, in figure 4.15, to just over $4\,\text{m/S}$ which is well below the velocity at which any wall instabilities might be expected in this material, even if there was no additional stabilising effect from the field gradient at the wall, and our assumption of a constant μ_w is reasonable. Figure 4.17 shows what happens if we continue our computer simulation to higher velocities. The relationship between v_x and x'

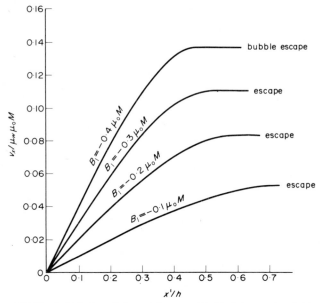

Figure 4.17. The bubble domain lags a distance x' behind the moving field well and this distance increases with velocity. Eventually $dv_x/dx' = 0$ and the bubble domain escapes completely.

becomes non-linear because R, in equation (4.34) is falling. At a critical velocity, dv_x/dx' becomes zero and the bubble domain escapes from the field well. It is interesting to note that this escape velocity can be quite low; for $B_1 = -0.1\,\mu_0 M$, it is only $11\,\text{m/S}$ for the $(\text{EuY})_3(\text{GaFe})_5\text{O}_{12}$ film considered above.

4.4   Conclusions

In this chapter we have reviewed the empirical data on bubble domain dynamics and then looked at the technical problem of producing stable bubble domain motion. This emphasised the importance of the stabilising field gradients at the wall of the moving bubble domain and the changes in size and shape which the bubble must undergo as a result of its motion.

Nothing has been said about the problem of understanding why a given material may show the particular wall mobility, μ_w, that it does and this problem is, of course, very important in deciding which material should be investigated. Hagedorn (1971), Vella-Coleiro (1972) and Bonner *et al.* (1972) have discussed the atomic scale energy loss mechanisms which are responsible for the mobility and the conclusion is that a very high mobility will be found in low loss materials in the sense that a very small local field will produce a finite wall velocity. Because the rate of energy loss from the moving wall can only be small in these low-loss materials, however, the wall structure will be forced to change as the velocity increases and instabilities of the kind described by Slonczewski (1971, 1972a, b, c, d) will occur more readily. It follows that the introduction of a well defined loss mechanism, like the substitution of Yb into the garnets described by Bonner *et al.* (1972), may improve the material characteristics from a bubble domain device point of view even though the low field wall mobility is reduced.

Having accepted a material, the dynamic behaviour of its bubble domains can be handled quite well numerically and the example given here of motion in a parabolic well shows that it is essential to maintain a high field gradient at the bubble wall if high velocities and good stability are to be obtained. Inertial effects, which come about because the bubble changes size as it moves, are also reduced by the use of high field gradients, as shown in figure 4.16.

With these results, we can now go on and consider the problem of magnetic overlays as bubble domain propagation tracks and see why it is that some kinds of overlay give a much better dynamic performance than others. We shall see that overlay patterns can be designed to produce a travelling magnetic field well very much like the simple one considered here in §§ 4.3.3 and 4.3.5, with the additional complication that the shape of the field well varies with time and also varies through the thickness of the bubble domain material. It follows that computer simulation of bubble domain motion in a practical overlay track is going to be much more complicated than the simple parabolic case considered here. The bubble domain will be changing size, not only because of changes in its velocity but also because of the varying shape of the field well. The same inertial effects which were discussed in § 3.4 will apply, however, and in the next chapter we shall consider how the field produced by the overlay pattern may be calculated so that, in principle, it should be possible to simulate bubble motion and find out how good a particular overlay track might be. The inertial effects described here, which are brought about by changes in the bubble domain diameter, will be important only if the velocity of the bubble changes during propagation.

References

Argyle, B. E., Slonczewski, J. C., and Mayadas, A. F., 1971, *AIP Conf. Proc.*, No. 5, 175.
Bitter, F., 1931, *Phys. Rev.*, **38**, 1903.
Bobeck, A. H., Fischer, R. F., Perneski, A. J., Remeika, J. P., and van Uitert, L. G., 1969, *IEEE Trans. Magnetics*, MAG 5, 544.
Bobeck, A. H., Danylchuk, I., Remeika, J. P., van Uitert, V. G., and Walters, E. M., 1970, *Proc. Int. Conf. on Ferrites*, Japan, July 1970, 361.
Bobeck, A. H., Blank, S. L., and Levinstein, H. J., 1972, *Bell Syst. Tech. J.*, **51**, 1431.
Bonner, W. A., Geusic, J. E., Smith, D. H., Rossol, F. C., van Uitert, L. G., and Vella-Coleiro, G. P., 1972, *J. appl. Phys.*, **43**, 3226.
Callen, H., and Josephs, R. M., 1971, *J. appl. Phys.*, **42**, 1977.
Callen, H., Josephs, R. M., Seitchik, J. A., and Stein, B. F., 1972, *App. Phys. Lett.*, **21**, 363.
Copeland, J. A., and Spiwak, R. R., 1971, *IEEE Trans. on Magnetics*, MAG 7, 748.
Döring, W. K., 1948, *Zeit. für Naturforschung*, **3a**, 373.
Fischer, R. F., 1971, *IEEE Trans. on Magnetics*, MAG 7, 741.
Galt, J. K., 1954, *Bell Syst. Tech. J.*, **33**, 1023.
Gyorgy, E. M., 1960, *J. appl. Phys.*, **31s**, 110s.
Gyorgy, E. M., 1963, Magnetisation reversal in non-metallic ferromagnets, in *Magnetism*, vol. III. Academic Press, ed. G. T. Rado and H. Suhl, 525–52.
Hagedorn, F. B., 1971, *AIP Conf. Proc.*, No. 5, p. 72.
Henry, G. R., 1971, *J. appl. Phys.*, **42**, 3150.
Höfelt, M. H. H., 1973, *J. appl. Phys.*, **44**, 414.
Kittel, C., and Galt, J. K., 1956, Ferromagnetic domain theory, in *Solid State Physics*, vol. 3. Academic Press, ed. F. Seitz and D. Turnbull, 437–564.
Landau, L. D., and Lifshitz, E., 1935, *Physikalische Zeitschrift der Sowjetunion*, **8**, 153.
Malozemoff, A. P., 1972, *Appl. Phys. Lett.*, **21**, 149.
Malozemoff, A. P., and Slonczewski, J. C., 1972, *Phys. Rev. Lett.*, **29**, 952.
Moore, G. E., 1971, *IEEE Trans. on Magnetics*, MAG 7, 751.
O'Dell, T. H., 1973, *Phil. Mag.*, **27**, 595.
Perneski, A. J., 1969, *IEEE Trans. on Magnetics*, MAG 5, 554.
Palmer, W., and Willoughby, R. A., 1967, *IBM J. of Res. and Dev.*, **11**, 284.
Rossol, F. C., 1969, *J. appl. Phys.*, **40**, 1082.
Rossol, F. C., and Thiele, A. A., 1970, *J. appl. Phys.*, **41**, 1163.
Rossol, F. C., 1971, *IEEE Trans. on Magnetics*, MAG 7, 142.
Rosencwaig, A., 1972, *Bell Syst. Tech. J.*, **51**, 1440.
Seitchik, J. A., Doyle, W. D., and Goldberg, G. K., 1971, *J. appl. Phys.*, **42**, 1272.
Shumate, P. W., 1971, *IEEE Trans. Magnetics*, MAG 7, 479.
Shumate, P. W., Smith, D. H., and Hagedorn, F. B., 1973, *J. appl. Phys.*, **44**, 449.
Slonczewski, J. C., 1971, *AIP Conf. Proc.*, No. 5, p. 170.
Slonczewski, J. C., 1972a, *Int. Jour. Mag.*, **2**, 85.
Slonczewski, J. C., 1972b, *IBM Res. Rep.*, RC3882, June 8th.
Slonczewski, J. C., 1972c, *IBM Res. Rep.*, RC3908, June 27th.
Slonczewski, J. C., 1972d, *IBM Res. Rep.*, RC4067, Oct. 3rd.
Slonczewski, J. C., and Malozemoff, A. P., 1972, *Proc. Conf. on Mag. and Mag. Mats.*, Denver, Nov. 1972 (*AIP Conf. Proc.* No. 10, p. 458).
Tabor, W. J., Bobeck, A. H., Vella-Coleiro, G. P., and Rosencwaig, A., 1972, *Bell Syst. Tech. J.*, **51**, 1427.
Thiele, A. A., 1971, *Bell Syst. Tech. J.*, **50**, 725.
Vella-Coleiro, G. P., 1972, *Proc. Conf. Mag. and Mag. Mats.*, Denver, Nov. 1972 (*AIP Conf. Proc.* No. 10, p. 424).
Vella-Coleiro, G. P., and Tabor, W. J., 1972, *Appl. Phys. Lett.*, **21**, 7.
Vella-Coleiro, G. P., Rosencwaig, A., and Tabor, W. J., 1972, *Phys. Rev. Lett.*, **29**, 949.
Vella-Coleiro, G. P., Hagadorn, F. B., Chen, Y. S., and Blank, S. L., 1973, *Appl. Phys. Lett.*, **22**, 324.
Wolfe, R., and North, J. C., 1972, *Bell Syst. Tech. J.*, **51**, 1436.
Yoshimi, K., Fujiwara, S., Yamauchi, F., and Furuoya, T., 1972, *IEEE Trans. Magnetics*, MAG 8, 669.

5 Bubble Domain Propagation Tracks and Detectors

In this chapter we shall consider the design of bubble domain propagation tracks and detectors for bubble domains. These topics are taken together because both involve a study of the external magnetic field of the bubble domain and its interaction with some kind of magnetic overlay pattern. Propagation tracks and detector patterns are both fabricated by very similar processes, in fact some designs make use of the same material to perform both functions simultaneously, so that the two subjects belong together for this reason as well.

It would appear that the design of bubble domain propagation tracks is the most difficult of all the steps taken in the design of a bubble domain device. This comes about because of the very large number of dimensional and material parameters which are involved and also because a difficult three dimensional magnetic field problem has to be solved. In practice, a considerable amount of computing must be done to optimise a design and this must be supported by accurate experimental data on the materials being used. The fabrication process also has to be very well defined so that dimensions, thicknesses and material constants may be known accurately.

To put these problems in perspective, we shall begin by considering the simplest situation where a magnetic field suitable for bubble propagation is to be generated by means of small bars of magnetic material laid close to the surface of the layer which supports the bubble domains. This is the most common technique in practice and was described qualitatively in Chapter 1, § 1.4.2. After this, we shall go on to review the patterns which have been used in practice and find some reasons why certain patterns would be expected to behave more effectively than others. The field calculations which are involved will then be used to consider the behaviour of the magneto-resistive detector, which is the most common form of bubble domain detector, and the chapter will conclude with a brief review of work which has been published on the fabrication problems of an integrated overlay and detector system.

5.1 The Problem of the Interaction Between a Bubble Domain and an Overlay Bar

5.1.1 *Experimental work*

The simplest experimental situation which may be set up to show the interaction between a bubble domain and an overlay bar is shown in figure 5.1. This photograph shows a number of isolated bars of evaporated 80/20 nickel/iron alloy, permalloy, spaced about 1 μm from the surface of an

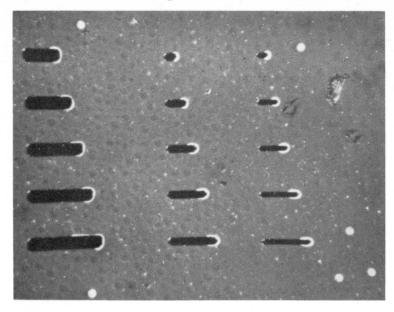

Figure 5.1. A Faraday rotation micrograph showing some isolated permalloy bars evaporated on top of an 8 μm layer of $(EuY)_3(GaFe)_5O_{12}$ garnet spaced off by a 1 μm silica film. The second bar down in the third column is 21 μm × 4·6 μm × 0·45 μm and is seen to support a circular bubble domain. Note that the bubble under the bar has a slightly larger diameter than the free bubbles which can be seen in this photograph (Jones, 1973).

epitaxial garnet layer. The geometry of the experiment is shown in figure 5.2 which illustrates the large number of variables involved. These are the permalloy bar dimensions (a, b, c), the bubble domain dimensions (D, h), the spacing of the bar from the garnet (s) and the location of the bubble relative to the bar $(d/2)$. On top of these we have the magnetic properties of the permalloy, the way in which the magnetisation of the permalloy varies with position and the effect of the magnetic field of the bubble domain itself. All these variables and parameters will come into the problem.

Some very interesting experimental work on this problem was published by George and Chen (1972) in which the bubble domain itself was used to measure the shape and intensity of the mean field under a single permalloy bar of the kind being discussed here. Measurements of this kind are very valuable as an introduction to the overlay problem and some very similar results are shown here in figure 5.3, which are due to Jones (1973), and apply for one of the permalloy bars shown in figure 5.1. A plot of bubble diameter against applied bias field for a free bubble and a bubble located at the end of a bar, as shown in figures 5.1 and 5.2, is given for the case of a constant in-plane field of 20 × 10^{-4} T. The difference between the two curves shown

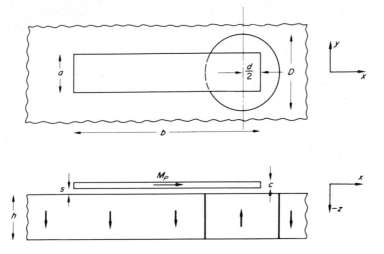

Figure 5.2. Defines the symbols used in the single overlay bar problem.

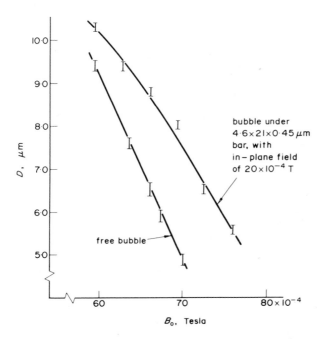

Figure 5.3. Experimental results showing the diameter of a bubble domain as a function of bias field when it is free and when it is located under an overlay bar (Jones, 1973).

118   *Magnetic Bubbles*

in figure 5.3 may then be taken to give the normal component of the magnetic field produced by the bar and this has been plotted as a function of
radius in figure 5.4. Over the range of *r* available in this experiment, limited
by elliptical run-out on the one hand and the disappearance of the bubble
domain under the bar on the other, we find the field under the bar to be of
the expected well shape. For this diagram we have taken the bias field, B_0,
to be positive so that the well field is negative. This is the same choice of
signs as that made for figure 1.14, where this problem was first introduced.

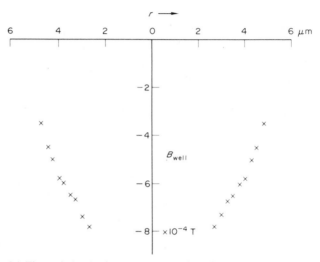

Figure 5.4. The variation in the *z* component of the field due to the bar referred to
in figure 5.3.

As the garnet layer used in this experiment was approximately 8 μm
thick, a typical bubble diameter for device applications would be 8 μm and
figure 5.4 shows that, for this bubble, the overlay is producing a field of
nearly 10 per cent of the applied bias field and that the gradient of the
overlay field is approximately 2×10^{-4} T/μm. It should be noted that this
is the same order of field gradient mentioned in Chapter 4, § 4.3.5, for the
higher values of inertial time constant.

Our object now is to formulate a theoretical model of the overlay bar
which will enable us to understand how the field which has been determined
experimentally comes about. Having done this, the problem will be looked
at from a more synthetic point of view in an attempt to see what the optimum
dimensions and spacing of the overlay bar might be.

5.1.2   *The magnetisation and magnetic field of an overlay bar*

Because the bubble domains in the experiment described above remained
circular when they were under the influence of the overlay bar field, as

shown in figure 5.1, we know that the field produced has circular symmetry. This was implied by figure 5.4, which gives the magnitude of the overlay bar field as a function of distance, r, from the bubble domain centre. The problem now is to find the magnetisation within the overlay bar which will be compatible with the applied fields and which will also generate the intensity and shape of the field found experimentally.

This is a very complex problem which can only be dealt with numerically and has been the subject of a number of papers. The important point to note initially is that the permalloy bar under discussion here is very far from being magnetically saturated. This was shown by Lin (1972), and discussed by Archer *et al.* (1972). Lin took Bitter patterns of the domain structure in overlay bars on bubble domain devices and the situation which was found is shown in figure 5.5 where the demagnetised state of a bar of

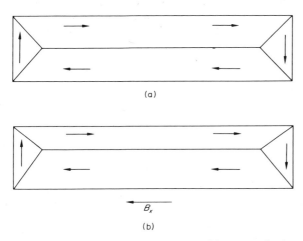

(a)

(b)

Figure 5.5. Typical magnetic domain structure in a thin rectangular bar when in zero field (a) and when in a field strong enough to produce 20 per cent saturation at the centre (b).

typical overlay dimensions is shown in figure 5.5(a). This kind of domain pattern in thin plates is well known and further experimental work may be found in papers by Gemperle (1966), De Blois (1968) and Jones (1970). The effect of an in-plane magnetic field, B_x, is shown in figure 5.5(b). This shows the overlay bar becoming magnetised to about 20 per cent of saturation at the centre which involves only a small movement of the central domain wall and a small change in the shape of the closure domains at either end of the bar. The component of magnetisation in the x direction falls smoothly to zero at the ends of the overlay bar under these conditions and it is precisely this kind of distribution of the magnetisation which will allow a symmetrical kind of field well, and thus a circular bubble domain, to exist under the end of the overlay bar.

Numerical analysis of the single overlay bar problem has been given by Copeland (1972) and George and Archer (1973a) and extended to more complex patterns by Lin (1972), Kempter (1972) and George and Archer (1973b). We shall consider the problem of the single overlay bar here by following the methods given by these authors and direct attention towards the synthetic, as opposed to the analytic, as much as possible because the questions to be answered are more concerned with what the dimensions associated with the overlay bar should be in order to provide the required field, rather than being concerned with what the field is once we are given the overlay dimensions.

5.1.3   *Calculating the magnetic field*

It was indicated in Chapter 2, § 2.1.3, that problems in magnetism may be considered using only the vectors B and M. In considering the problem of the magnetic field of an overlay bar, however, we shall introduce the third magnetic vector, H, which is related to B and M by the entirely general relationship

$$B = \mu_0(H + M) \tag{5.1}$$

and H will be referred to as the magnetic excitation. The introduction of H in this problem will allow a straightforward comparison between the results given here and the results of the references given above.

If we know what form the magnetisation takes within a body, the form of H may be calculated directly by means of the well-known result

$$H = (1/4\pi)\,\mathrm{grad}\left\{ \int_v \frac{\mathrm{div}\,M\,\mathrm{d}v}{r_{ik}} + \int_s \frac{M \cdot n\,\mathrm{d}s}{r_{ik}} \right\} \tag{5.2}$$

The most concise derivation of this result may be found in Sommerfeld (1952, p. 78) and a detailed consideration of the mathematical method which is used may be found in Sommerfeld (1950, p. 147). The first integral in equation (5.2) is taken over the entire volume of the magnetised body, r_{ik} being the distance between the point of integration, i, which is surrounded by the volume element $\mathrm{d}v$, and the point, k, at which H is to be evaluated. The second integral in equation (5.2) is taken over the entire surface of the magnetised body, n being a unit vector normal to this surface at the point of integration and directed into the body. Having found H, by means of equation (5.2), the magnetic field B follows at once from equation (5.1) because M has already been defined.

A particularly simple example is given in figure 5.6(a) where a bar of typical overlay shape, $a:b:c = 1:5:\cdot05$, using the symbolism of figure 5.2, is treated as being uniformly magnetised to some level $\hat{M}$ which is less than, or equal to, the saturation magnetisation of the overlay material, M_p. In this case M, in equation (5.2), is simply $M_x = \hat{M}$, $M_y = M_z = 0$, and $\mathrm{div}\,M$

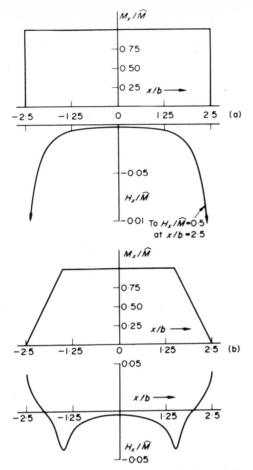

Figure 5.6. The magnetic excitation, H_x, in an overlay bar of dimensions $a:b:c$ = $1:5:\cdot05$. Case (a) shows the very high value of H_x when the bar is saturated. When the magnetisation falls off smoothly at either end, case (b), H_x is reduced considerably.

is zero. The only non-zero term in equation (5.2) is the second integral and, evaluating this, we find that the x component of $\boldsymbol{H}$ along the axis of the bar, which is the x axis, in figure 5.2, has the form shown as H_x in figure 5.6(a). H_x is very small compared to $\hat{M}$ at the centre of the bar but rises to a value $-\hat{M}/2$ at the plane ends.

This is why we do not find uniformly magnetised bodies, except under very special conditions. Because the magnetic energy density, $-\frac{1}{2}\boldsymbol{B}.\boldsymbol{H}$, is equal to $-(H^2 + \boldsymbol{M}.\boldsymbol{H})\mu_0/2$, in view of equation (5.1), there will be a very high positive energy density associated with the ends of the uniformly magnetised bar of the order of $\mu_0\hat{M}^2/8$. A very large applied field, of the order of $\mu_0\hat{M}/2$, would be required to reduce this and support this uniform state of magnetisation.

If we now refer back to figure 5.5, and ask what kind of distribution of magnetisation would actually be found in practice, it is clear that the true situation in an overlay bar is going to be more like that shown in figure 5.6(b). Here, the central part of the overlay bar is shown magnetised to $\hat{M}$, in the x direction, while M_x falls off to zero at either end. Such a distribution corresponds to the manner in which M_x varies in figure 5.5. The closure domains shown in figure 5.5 are magnetised in the y direction and may be ignored with little error. If equation (5.2) is now applied to the situation shown in figure 5.6(b), it is the second integral in equation (5.2) which vanishes this time, because $\boldsymbol{M} \cdot \boldsymbol{n}$ is zero everywhere on the surface, and it is the first integral which must be considered. The value of div $\boldsymbol{M}$ is $\hat{M}/d$, within the regions at the end of the bar where $\boldsymbol{M}$ is changing, and the result for H_x is shown in figure 5.6(b). The numerical integration used to obtain this result has rounded off the variation in M_x with x to the order of the dimension c, in figure 5.2, otherwise dM_x/dx would be discontinuous. This allows a sensible evaluation of H_x at the points $x = \pm b$ and $\pm(b - d)$. The result shows that H_x remains small within the bar and that this kind of distribution of magnetisation is compatible with a small applied field.

We have avoided going into the detailed algebra of the problem at this stage because this will be necessary for a related problem which is considered in the next section, and in § 5.1.7 where we shall have to do some detailed calculations concerning the applied field and the field of the bubble domain itself. These are analytic problems and our purpose in postponing them is that we do not yet know what it is we need from the overlay bar. To find out what we can about this we now turn our attention to the external field of the overlay bar, this being the field which is used to control the bubble domain in the device.

5.1.4   *The external magnetic field of an overlay bar*

We shall now calculate the external magnetic field of an overlay bar which is magnetised in the manner shown in figures 5.5(b) and 5.6(b). The situation is that shown in figure 5.7 where the field is to be evaluated at the point k, outside the bar and equation (5.1) is simply

$$\boldsymbol{B} = \mu_0 \boldsymbol{H} \tag{5.3}$$

because $\boldsymbol{M}$ is zero outside the overlay bar. We are concerned with the z component of $\boldsymbol{B}$ at k, because this is the field component which subtracts from the applied bias field in the device and forms the field well for the location of the bubble domain. Equation (5.2) then simplifies to

$$B_z = \frac{\mu_0}{4\pi} \frac{\mathrm{d}}{\mathrm{d}z} \int_{-a/2}^{+a/2} \int_{-c/2}^{+c/2} \left\{ \int_{-b/2}^{-b/2+d} \frac{\hat{M}/d}{r_{i'k}} \, \mathrm{d}x' + \int_{b/2-d}^{b/2} \frac{-\hat{M}/d}{r_{ik}} \, \mathrm{d}x' \right\} \mathrm{d}z' \, \mathrm{d}y' \tag{5.4}$$

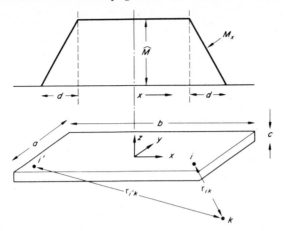

Figure 5.7. Defines the symbolism of equation (5.4). The points of integration are i' and i while the field is evaluated at point k.

where $k = (x, y, z)$, $i = (x', y', z')$ and $i' = (-x', y', z')$ in the (x, y, z) system shown in figure 5.7. Under these definitions, r_{ik} and $r_{i'k}$ are given by the same expression

$$r_{ik} = [(x - x')^2 + (y - y')^2 + (z - z')^2]^{1/2} \tag{5.5}$$

We now differentiate under the integral sign and integrate with respect to z' by using the identity

$$\int \frac{u \, du}{(u^2 + a^2)^{3/2}} = \frac{-1}{(u^2 + a^2)^{1/2}} \tag{5.6}$$

where a is a constant with respect to the variable u, and then integrate again, with respect to x', by using the identity

$$\int \frac{du}{(u^2 + a^2)^{1/2}} = \log [u + (u^2 + a^2)^{1/2}] \tag{5.7}$$

The final result is that

$$\frac{B_z}{\mu_0 \hat{M}} = \frac{1}{4\pi d} \int_{-a/2}^{+a/2} \log_e \left(\frac{R_1 R_3 R_5 R_7}{R_2 R_4 R_6 R_8} \right) dy' \tag{5.8}$$

where,

$$R_1 = (b/2 - x) + [(b/2 - x)^2 + (y' - y)^2 + (z - c/2)^2]^{1/2} \tag{5.9}$$

$$R_2 = (b/2 - x) + [(b/2 - x)^2 + (y' - y)^2 + (z + c/2)^2]^{1/2} \tag{5.10}$$

$$R_3 = (b/2 - d - x) + [(b/2 - d - x)^2 + (y' - y)^2 + (z + c/2)^2]^{1/2} \tag{5.11}$$

$$R_4 = (b/2 - d - x) + [(b/2 - d - x)^2 + (y' - y)^2 + (z - c/2)]^{1/2} \tag{5.12}$$

$$R_5 = (-b/2 + d - x) + [(-b/2 + d - x)^2 + (y' - y)^2 + (z + c/2)^2]^{1/2} \quad (5.13)$$

$$R_6 = (-b/2 + d - x) + [(-b/2 + d - x)^2 + (y' - y)^2 + (z - c/2)^2]^{1/2} \quad (5.14)$$

$$R_7 = (-b/2 - x) + [(-b/2 - x)^2 + (y' - y)^2 + (z - \dot{c}/2)^2]^{1/2} \quad (5.15)$$

$$R_8 = (-b/2 - x) + [(-b/2 - x)^2 + (y' - y)^2 + (z + c/2)^2]^{1/2} \quad (5.16)$$

Equation (5.8) may be evaluated numerically because none of the R expressions above become zero when the point k is outside the bar.

5.1.5 *The magnetisation required for a cylindrically symmetric field well*

One striking feature of the experimental study of the single overlay bar, which was discussed in §§ 5.1.1 and 5.1.2, is the observation that the bubble domain remains circular when it is located under one end of the bar. This means that the field well produced by the bar has cylindrical symmetry, a point implied by the data presented in figure 5.4. A further observation, which was not mentioned in §5.1.1, is that the location of the bubble centre, $d/2$ in figure 5.2, is such that $d/2 \approx a/2$ for a wide range of in-plane fields of the order of intensity which would be found in a bubble domain device.

The conclusion to be drawn from this is that our first application of the equations derived in § 5.1.4 should be to seek a value of d which will provide a field well with cylindrical symmetry. The use of the same symbol, d, in figures 5.2 and 5.7 anticipates the result of this search. We find that the centre of the field well produced by the overlay bar does occur very close to $x = (b - d)/2$ and that cylindrical symmetry occurs when $d \approx a$. This result is quickly obtained by using the computer to evaluate equation (5.8) in an interactive mode where a plot of B_z is made in both the x and y directions and d is varied until the two curves appear to be the same shape.

Results for an overlay bar of the dimensions used in the experiment, for which figures 5.3 and 5.4 apply, are shown in figures 5.8 and 5.9. It may be seen that the two curves are very nearly identical in the region 3 μm to 5 μm from the field well centre and it is in this region that we are able to superimpose the experimental results from figure 5.4 by assigning a value to $\mu_0 \hat{M}$. Knowing the saturation magnetisation for permalloy is such that $\mu_0 M_p = 1 \cdot 0$ T, we find that $\hat{M}/M_p = 19 \cdot 3$ per cent for the fit shown in figure 5.8 and 20·6 per cent for that shown in figure 5.9. We conclude that the overlay bar in this experiment is about 20 per cent saturated by the combined influence of the in-plane field and the bubble domain stray field. The bubble domain diameter chosen for this fit was 8 μm because this was the intended device diameter in this particular case, the garnet layer being approximately 8 μm thick.

5.1.6 *The variation in the magnetic field with depth*

Figures 5.8 and 5.9 show the z component of the magnetic field due to the overlay bar at the centre of the 8 μm thick garnet layer which is being con-

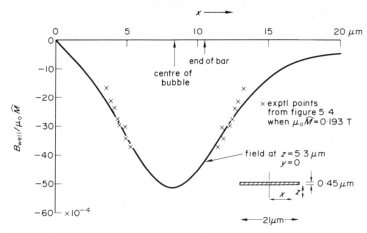

Figure 5.8. The z component of the field under the overlay bar predicted by equation (5.4). The experimental points from figure 5.4 are also shown. The dimensions are $a = d = 4·6\ \mu m$, $b = 21\ \mu m$ and $c = 0·45\ \mu m$.

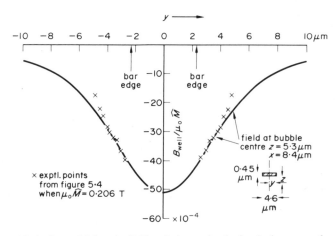

Figure 5.9. As figure 5.8, but the field variation under the bar is shown as a function of y.

sidered in this particular example. This field varies quite strongly with z and, for the case of an 8 μm diameter bubble, this variation is shown in figure 5.10. It so happens that the mean value of this field is very close to the value it has at the centre, $z = 5·3\ \mu m$, and this is why the central field shown in figures 5.8 and 5.9 was used to compare these theoretical results with the experimentally determined mean field shown in figure 5.4. Although the field variation shown in figure 5.10 looks quite severe, we must remember that this overlay bar field is superimposed upon a bias field of $\approx 68 \times 10^{-4}$ T, as shown in figure 5.3, and its mean value is only $5·8 \times 10^{-4}$ T, as shown in

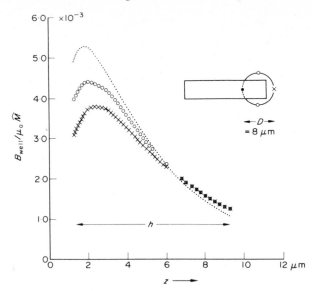

Figure 5.10. The z component of the field under the overlay bar as a function of z at the points on the periphery of the 8 μm diameter bubble domain shown.

figure 5.4. No measurable distortion of the bubble domain would be expected as a result of such a small variation.

There is one aspect of this variation in the field with depth which may be important and this concerns the radial gradient of the field. Figure 5.11 shows what is meant by this by repeating the right hand side of figure 5.9, showing how the well field varies with y, and including the results for the top and the bottom of the garnet layer. The results showing the variation with x are very similar and show, as does figure 5.11, that the radial field gradient is very much greater at the top of the bubble domain than at the bottom. For the 8 μm diameter bubble domain, being used as an example here, we find that the gradient, $\partial B_z/\partial r$, at the top, centre and bottom of the bubble wall is varying as 15:3:1. Referring to Chapter 4, § 4.3.5, it is clear that this means that we should expect the dynamic behaviour of a bubble domain in the kind of field well shown in figure 5.11 to be predominantly determined by the field near the top of the bubble, not just because this field is larger by a few per cent but because its radial gradient is considerably larger.

Figure 5.11 also shows that there is a bubble domain radius, around 5 μm, where the field gradient does not vary anywhere near so much as it does at 4 μm radius. The variation in the field with z, for this 10 μm diameter bubble, is much smaller too and this is shown in figure 5.12. The conclusion is that better performance might be expected with an 8 μm diameter bubble if the width of the overlay bar was reduced.

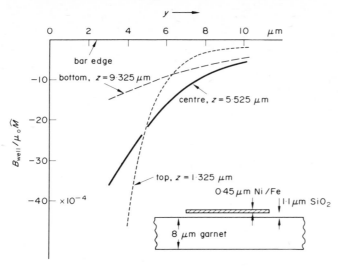

Figure 5.11. The z component of the field under the overlay bar is shown as a function of y for three values of z; these defining the top, centre and bottom of the garnet layer.

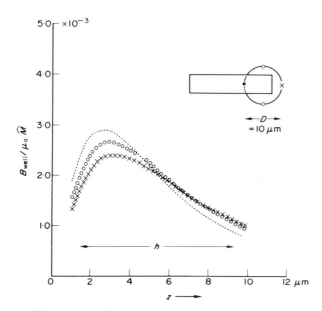

Figure 5.12. As figure 5.10, but for a larger diameter bubble.

5.1.7 *The magnitude of the in-plane field*

We now have to calculate what magnitude of in-plane field, or drive field, will be required to support the magnetisation shown in figure 5.6(b). It is this particular distribution of M_x which we have shown will produce the symmetrical field well which is found experimentally and we have also shown, in § 5.1.5 that $\hat{M}/M_p$ must be of the order of 20 per cent in order to explain the intensity of the well field produced by this particular bar.

The in-plane magnetic field which is needed to support this distribution of M_x comes from two separate sources. The first is the uniform applied drive field, B_D, and the second is the stray field of the bubble domain under the bar. We shall consider the stray field of the bubble domain first and, to do this, we shall look at the particular problem of the $4.6 \times 21 \times 0.45\,\mu m$ permalloy bar which has been the subject of our previous calculations.

Figure 5.13 shows the stray field of the bubble taken along the x axis of the permalloy bar when the bubble domain is located $8.4\,\mu m$ from the centre of the bar, this being the centre of the field well shown in figure 5.8, and is $8\,\mu m$ in diameter. The value of $\mu_0 M$ in the material which was used for these experiments was $136 \times 10^{-4}\,T$ and this value has been used to obtain the absolute value of the field.

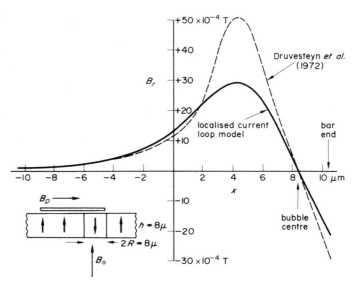

Figure 5.13. The radial component of the stray field of an $8\,\mu m$ bubble domain evaluated in the plane of the overlay bar. The bubble field aids the applied drive field, B_D, over most of the bar length.

Two quite distinct models have been used to calculate this field. The higher values are obtained by interpolation from the tables given by Druvesteyn *et al.* (1972) who calculated the stray field as that from a saturated cylinder. Similar results were given by Goldstein and Copeland (1971). This model is very sensitive to the exact shape and width of the domain wall where it passes underneath the overlay bar. For the case of a zero width wall and a perfectly saturated bubble domain, an infinite field is predicted by this model immediately above the domain wall and this infinity is entered in the tables given by Druvesteyn *et al.* (1972) despite the fact that such a result is not possible in the real world. The effect of this singularity in the solution is to give an unrealistic maximum to the field.

A model which underestimates the field at this point is one which represents the bubble domain by means of a circular current loop, radius R, placed at the centre of the garnet layer. As in Chapter 2, § 2.3.1, the current flowing in this loop is $2Mh$. The external field of such a current loop is a well known problem and the radial component of this field is given by the expression,

$$B_r = \frac{\mu_0 I}{4\pi}\left(\frac{1}{R}\right)\left(\frac{z}{R}\right)\left(\frac{R}{r}\right)^{3/2} k\left[\frac{(2 - k^2)}{2(1 - k^2)}E(k) - K(k)\right] \qquad (5.17)$$

where

$$k^2 = \frac{4(r/R)}{(r/R + 1)^2 + (z/R)^2} \qquad (5.18)$$

is the argument of the first and second kind complete elliptic integrals $K(k)$ and $E(k)$ (Jahnke and Emde 1945). Equation (5.17) is a slight rearrangement of the form of the solution given by Smythe (1968). The field, B_r, is evaluated at the point (z, r) in cylindrical coordinates with the centre of the coil at the origin, z being the axis of the coil. To represent the bubble domain we replace I, in equation (5.17), by $2Mh$, the equivalent domain wall current discussed in Chapter 2, § 2.3.1.

Figure 5.13 shows that the two models give the same result at distances well away from the domain. Close to the domain wall, the localised current loop underestimates the field but it would be nearer the truth than the alternative model. It may be seen that the field due to the bubble domain is very non-uniform and it is acting mainly in the same direction as the applied in-plane field, which was 20×10^{-4} T in this particular experiment. The mean value of the resulting total field is $29 \cdot 8 \times 10^{-4}$ T.

We shall now look for a simple method of estimating the required in-plane field. This is possible if we look back to figure 5.6 and the discussion of § 5.1.3, where the magnetic excitation H_x was calculated for the kind of magnetisation distribution which we are dealing with here. The details of the calculation which led to the curves for H_x in figure 5.6 was not given in § 5.1.3 but this problem is almost identical to the one considered in § 5.1.4, except that equation (5.4) involves the operation d/dx, equation (5.6) is used to perform the integration with respect to x' and equation (5.7) is used to perform the integration with respect to y'. The final result is then

$$\frac{H_x}{\hat{M}} = \frac{1}{4\pi d} \int_{-a/2}^{+a/2} \log_e \left\{ \frac{S_1 S_3 S_5 S_7}{S_2 S_4 S_6 S_8} \right\} dy' \qquad (5.19)$$

corresponding to equation (5.8), where

$$S_1 = (c/2 - z) + [(b/2 - x)^2 + (y' - y)^2 + (c/2 - z)]^{1/2} \qquad (5.20)$$

$$S_2 = -(c/2 + z) + [(b/2 - x)^2 + (y' - y)^2 + (c/2 + z)]^{1/2} \qquad (5.21)$$

$$S_3 = -(c/2 + z) + [(b/2 - d - x)^2 + (y' - y)^2 + (c/2 + z)^2]^{1/2} \qquad (5.22)$$

$$S_4 = (c/2 - z) + [(b/2 - d - x)^2 + (y' - y)^2 + (c/2 - z)^2]^{1/2} \qquad (5.23)$$

$$S_5 = -(c/2 + z) + [(-b/2 + d - x)^2 + (y' - y)^2 + (c/2 + z)^2]^{1/2} \qquad (5.24)$$

$$S_6 = (c/2 - z) + [(-b/2 + d - x)^2 + (y' - y)^2 + (c/2 - z)^2]^{1/2} \qquad (5.25)$$

$$S_7 = (c/2 - z) + [(-b/2 - x)^2 + (y' - y)^2 + (c/2 - z)^2]^{1/2} \qquad (5.26)$$

$$S_8 = -(c/2 + z) + [(-b/2 - x)^2 + (y' - y)^2 + (c/2 + z)^2]^{1/2} \qquad (5.27)$$

Equation (5.19) is then evaluated numerically and the result for the particular overlay bar under discussion is shown in figure 5.14. The discontinuities in dM_x/dx are smoothed out in the numerical integration, because M_x would be analytic in nature, so that reasonably accurate extreme values of H_x may be evaluated. These extreme values do not affect the mean value of the function $H_x(x)$ to any considerable extent and, we shall see, it is the mean value we are mainly concerned with.

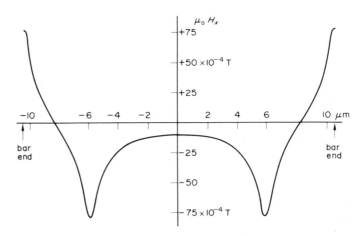

Figure 5.14. The so called demagnetising field $\mu_0 H_x$ of the overlay bar along the x axis. The dimensions are $a = d = 4.6\ \mu m$, $b = 21\ \mu m$ and $c = 0.45\ \mu m$. The value of $\hat{M}$, figure 5.7, is $0.2\ M_p$, where $\mu_0 M_p$, for permalloy, is 1.0 T.

We now have enough information to find out what order of in-plane field must be applied to support the magnetisation in the overlay bar. This is done by considering the energy density $-\frac{1}{2}\boldsymbol{B}.\boldsymbol{H}$, as we mentioned in § 5.1.3. Here, $\boldsymbol{H}$ is the total magnetic excitation, that is H_x from equation (5.19). B_r/μ_0 from equation (5.17) and B_D/μ_0, where B_D is the uniform applied in-plane drive field. In view of equation (5.1), the energy density is given by $-(H^2 + \boldsymbol{M}.\boldsymbol{H})\mu_0/2$ and we note that $H \ll M$ in all practical cases. An approximation to the total energy density per unit cross-sectional area is thus

$$E = \frac{\mu_0}{2} \int_{-b/2}^{-b/2} M_x(H_x + B_r/\mu_0 + B_D/\mu_0)\,\mathrm{d}x \qquad (5.28)$$

The solution to the problem is as follows—Given M_x we have determined H_x and given the parameters and position of the bubble domain we may determine B_r/μ_0. These are the results shown in figures 5.14 and 5.13. We substitute M_x, B_r/μ_0 and H_x into equation (5.28) and solve for B_D/μ_0 with E set equal to zero. It is, of course, an approximation to say E is zero but the point is that E is small in all practical situations. The solution for B_D in this particular example is $B_D = 14.6 \times 10^{-4}$ T whereas the actual value used during measurement was 20×10^{-4} T. This is adequate accuracy and has allowed us to use very simple mathematical techniques which permit a synthetic approach to the problem. This is discussed below.

5.1.8   *Optimisation of an overlay bar*

The theory given in the preceding sections has allowed us to determine what kind of overlay bar magnetisation will produce a cylindrically symmetric field well, to calculate the degree of magnetisation, $\hat{M}/M_p$, to calculate the field of the bubble domain and, finally, to decide what the in-plane drive field should be in order to support this situation in the overlay bar. This has been done by means of a very simple and approximate method but, because of this, it has been possible to keep the physical situation in the foreground so that we can see what the important parameters are. Our example has only considered one bar and the important problem is how to move a bubble domain from one bar to another. This more difficult problem may also be approached using the theory given here.

To conclude this section, let us look back at the results which have been obtained and see how they affect the optimisation of the overlay pattern.

(i)   *Overlay bar thickness*   Equations (5.8) and (5.19) show that both B_z, the drive field produced by the overlay, and $\mu_0 H_x$, the so-called demagnetising field, are both very nearly proportional to the bar thickness c. This is because c is very much smaller than a or b in any typical overlay bar. This results in c being a very unimportant parameter because, given some in-plane drive field, the value of $\hat{M}/M_p$ in the overlay bar can adjust itself to accommodate different values of c. Figure 5.5 makes this particularly clear where figure 5.5(b) is in fact drawn to show the situation of $\hat{M} = 20$ per cent M. If the thickness of this overlay bar was doubled the central wall shown in figure 5.5(b) would simply move slightly towards the centre to make $\hat{M} \approx 10$ per cent M and the internal and external fields of the bar would be very nearly as before.

This is not true if the bar is made so thin that it has to saturate at its centre. In the example considered here a thickness below $0 \cdot 1 \ \mu m$ would require saturation if the external field, B_z, produced by the bar was to be the same as before.

(ii) *Overlay bar width and length* These are decided more by the propagation pattern itself and the use of bar shape to produce a travelling magnetic wave when a rotating field is applied. This is discussed in § 5.2, where we shall see that the patterns which use bars are exploiting the shape anisotropy of these bars so that the ratio of length to width will always be greater than unity. Typically we find $b/a \approx 4$ or 5.

The important point is how wide a bar should be for a given bubble diameter and this question may be answered by referring to the results shown in figures 5.8, 5.9 and 5.11. The bar width should be selected to give maximum mean field gradient at the bubble wall, in view of the results given in Chapter 4, § 4.3.5. This decision is also affected by the bar to garnet spacing.

(iii) *Bar to garnet spacing* This is one of the most important parameters and is shown up when computations of the kind leading to figures 5.10, 5.11 and 5.12 are made. When the bar width is only slightly smaller than the bubble diameter, figure 5.11 shows that very large fields may be produced at the top of the bubble domain and the spacing, s, may have to be quite a considerable fraction of the garnet thickness. During propagation the bubble domain wall must pass near the edge of the overlay bar so that, even when the bar width is small, compared to the bubble diameter, a good spacing is essential. George and Archer (1973a) point out this interdependence of bar width and spacing in their very detailed study of overlay optimisation which was later extended to consider the combinations of overlay pattern elements (George and Archer 1973b).

5.2 Overlay Propagation Tracks

In the previous section we looked at the problem of a single bar overlay and deduced the form of its magnetisation, its internal and external magnetic field and the magnitude of the in-plane field required to support the magnetisation. We shall now look at the problem of making a propagation track by means of these overlay bars, and the more complex shapes which are derived from them.

The word 'overlay' is quite important here because we are discussing propagation tracks which are formed by means of a pattern of magnetic material which is formed on one side of the bubble domain material only. We shall see, in § 5.2.3, that there are other possibilities which may become of great technical importance in the future. The first useful bubble domain devices made use of the epitaxial garnet layers which were grown by the liquid phase epitaxy method, described in Chapter 3, and this limited the implementation of propagation tracks to the overlay variety.

5.2.1 *Forms of overlay track*

Some of the well known patterns used for bubble domain propagation tracks are shown in figure 5.15. It is clear from this that these are all derived from

Figure 5.15. Well known bubble domain propagation tracks. The bubble positions refer to the directions of the rotating in-plane field shown at (e).

the bar element and the principle of operation is that we are using the shape anisotropy of the bar and a rotating in-plane magnetic field to select bars sequentially and so produce a travelling magnetic wave for bubble propagation.

By shape anisotropy we mean that the state of magnetisation in a bar is changed considerably when a magnetic field is applied along its long axis. Little change in magnetisation results when an in-plane field is applied at right angles to this long axis. Consider the **TI** bar pattern of figure 5.15(a) as an example. The direction of the rotating field is shown at the bottom of the diagram. A transition or propagation of the $3 \rightarrow 4 \rightarrow 5$ variety is easily treated as a single bar with little error so that the form of the field locating the bubble at positions 3 and 5 is very similar to the one we have already calculated. At position 4 we have to consider two overlay bars at right angles, add the external fields and decide upon the optimum spacing between the **T** and the **I**.

What is much more difficult to calculate with the **TI** pattern is what is really happening in a propagation transition of the $1 \rightarrow 2 \rightarrow 3$ variety because the magnetisation within the head of the **T** bar must take up quite a complicated form as the bubble domain slides underneath it. This is a difficulty with the patterns shown in figures 5.15(a), (b) and (c). In all these patterns the bubble domain has to pass under the permalloy completely.

Figure 5.15(b) is called the **YI** pattern (Danylchuk 1971) and has been the subject of an interesting dynamic study experimentally by Fischer (1971) who pointed out a very important aspect of the rotating in-plane field. This was that there is no reason to restrict the rotating drive field to one of constant amplitude, as it has been shown in figure 5.15. Further flexibility may be introduced into the design by amplitude modulation; for example, some advantage would be intuitively expected if the amplitude was increased for positions 2, 4, 6 and 8 in the case of the **TI** pattern. Fischer's experiments allowed this because the rotating field was provided by two orthogonal coils which were supplied with square wave currents, 90° out of phase with one another. A rotating field of constant amplitude requires the simpler sinusoidal drive current.

Figure 5.15(c) shows the chevron propagation pattern (Bobeck *et al.* 1971a) which is similar to the **YY** pattern described by Yamauchi *et al.* (1972). George and Archer (1973b) made a thorough analysis of the **TI**, **YI** and chevron patterns and their results showed that the **TI** pattern produced a good travelling magnetic wave for bubble propagation in that the amplitude and gradient of the field well locating the bubble domain remained reasonably constant during the drive field cycle. The chevron pattern showed a rather poor field well, of small depth and gradient, for positions 7 and 3 in figure 5.15(c). The **Y** bar showed a considerable increase in gradient as the bubble moved to position 1 in figure 5.15(b) which suggested that the small tab on the **Y** bar was not a good idea and that it would be better to remove it thus giving a chevron with an angle different to the 90° shown in figure 5.15(c). Optimisation of the chevron angle has not been published so far but it is probably around 120°.

Figure 5.15(d) shows the **X** bar pattern described by Parzefall and Littwin (1973) and should represent an improvement over the other patterns because this is the only pattern shown in which the bubble domain does not have to pass under the overlay bar but runs from tip to tip. A deeper field well with a higher field gradient should be obtained from this pattern and Parzefall and Littwin reported operation at 600 KHz with this pattern compared to 300 KHz with a **YI** bar pattern on the same material.

5.2.2 *Other considerations in overlay tracks*

Figure 5.16 shows closed loop propagation tracks in the four patterns which were shown in figure 5.15. This illustrates a further factor which

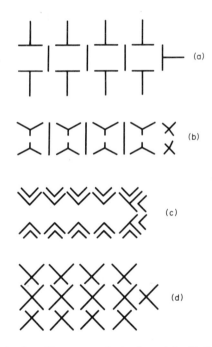

Figure 5.16. Showing how *U* turns may be implemented with the tracks shown in figure 5.15. Note that the chevron tracks (c) may consist of stacked chevrons. This has advantages in track to track transfers and interactions (Bobeck *et al.*, 1973).

decides the choice of a pattern and this is the area which it occupies. The **YI** pattern is particularly economical in this respect and this is why it is used for work with sub-micron bubbles where the object is to increase the bit density as much as possible (Hu *et al.* 1973).

The dynamic behaviour of propagation tracks is of considerable importance when high speed operation is required and some very elegant experimental techniques have been described by Rossol (1971) who applied stroboscopic techniques so that the motion of bubble domains along a propagation track could actually be observed. These experiments revealed considerable changes in bubble domain shape and diameter during propagation along a **TI** track and could be used to deduce the shape and intensity of the field produced by the overlay bars under dynamic conditions, a particular problem being the poor dynamic performance of the **TI** pattern corner illustrated in figure 5.16(a). A dynamic study of the **YI** bar pattern by Fischer (1971), which was mentioned above, was concerned with the problem of driving a single bubble domain along a propagation track for just a few periods of the rotating drive field and observing how rapidly this could be done until propagation failed. A great deal of work remains to be done on the dynamics of bubble domains within the propagation track environment.

A further factor in the choice of a propagation track is the ease with which crossovers, interactions between bubbles on neighbouring tracks and transfers of bubble domains from one track to another may be made. These problems have been discussed by Yamauchi *et al.* (1972) and by Bobeck *et al.* (1971, 1973) and examples may be found in description of complete devices (Chang *et al.* 1972, Bosch *et al.* 1973a, 1973b).

5.2.3 *Double sided and special propagation tracks*

The earliest bubble domain devices did not use epitaxial garnet films but polished slices of bulk crystals. With the orthoferrite crystals, described in Chapter 3, these single crystal slices were quite thick and could be handled without any form of support, so that it was natural to consider using a magnetic overlay pattern on both sides of the crystal.

An example of such a double sided propagation track is shown in figure 5.17 and was studied dynamically by Fischer (1971). The overlay bar field variation with z, discussed here in connection with figures 5.10 and 5.12, is smoothed out to a considerable extent because the bubble domain is alternately operated upon from the top and then the bottom. Dynamic performance is thus considerably better with these double sided tracks, not only for this reason but also because the track shown in figure 5.17, like the **X** bar track of figure 5.15(d), does not involve any complex magnetisation pattern to be produced in the bars. The predominant mode of magnetisation is that of a single bar magnetised along its length and alternating in direction. This alternation simply involves a small motion of the central domain wall shown in figure 5.5(a). The advantage of double sided overlay patterns has also been discussed by Chang *et al.* (1972) in connection with the bubble domain duplicator or splitter which was referred to in Chapter 1, § 1.4.4.

A number of very original and interesting propagation tracks of a rather special nature should be mentioned briefly. These may become of great

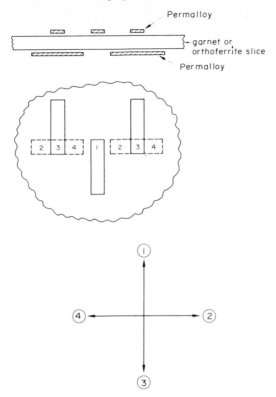

Figure 5.17. A double sided propagation track. Bubble domain positions are numbered according to the in-plane field directions shown at the bottom of the diagram.

importance because the idea is directed towards avoiding the difficult etching process step needed for the permalloy bar patterns which we have discussed up to now.

The first of these special propagation tracks to be considered involves the use of the bubble domain material itself as a magnetic overlay. Because the bubble domain material is magnetic, but has its easy direction of magnetisation normal to the plane of the layer, the problem is one of treating the surface of the layer in such a manner as to destroy this easy direction and cause the magnetisation to lie in-plane. Wolfe *et al.* (1972) showed that this could be done by means of ion implantation at the surface, which relieved the strain responsible for the magnetic anisotropy of some epitaxially grown garnet layers. An 'overlay' pattern could then be formed which was made of the garnet itself and which could be made to work in very much the same way as the permalloy overlays described above. A disadvantage is that we are now dealing with an 'overlay' material with a very much lower saturation magnetisation than permalloy so that, to get the same drive fields on the bubble domains, we need to take the 'overlay' regions much nearer to magnetic saturation. (Johnson *et al.* 1973).

A similar idea to ion implantation has been described by Le Craw *et al.* (1973). This involves changing the saturation magnetisation of the bubble domain layer in selected areas by the diffusion of silicon into the garnet material. The value of M could be reduced by this silicon diffusion to nearly one third of its previous value and this allowed confining tracks to be formed within the garnet layer. Propagation was achieved by current carrying conductors or by ratchet like patterns which responded to pulsating in-plane field gradients.

A further group of special propagation tracks respond to a pulsating magnetic field which is added to the bias field, instead of being in-plane as is the case with all the field access propagation tracks we have discussed up to now. These are the 'angelfish' variety, based upon a track first described by Bobeck *et al.* (1969) which used small triangles of permalloy overlay in a nose to tail configuration. The pulsating bias field caused the bubble domains in this track to expand and contract, ratcheting their way along the track at the same time. Improvements in this kind of track were made by Hayashi *et al.* (1972a) who used a sawtoothed permalloy rail guide in place of the angelfish pattern, this having the advantage that the bubble domains were not obscured by the track and could be used to generate a display. A similar track using triangular depressions etched into the surface of the bubble domain material was described by Hayashi *et al.* (1972b).

5.2.4   *Development towards the optimum propagation track*

To conclude this section on propagation tracks for bubble domain devices we shall briefly consider the directions development could go in order to reach an optimum solution to this problem.

Propagation alone is not sufficient for a bubble domain device, the function of generation and interaction, reviewed in Chapter 1, §§ 1.4.2 and 1.4.4, are also essential. For a given variety of overlay pattern, we find that all these processes may be implemented but when we look at the range of bias fields and in-plane drive fields which allow all these processes to happen together, it may well turn out that the operating margins involved are too narrow for a practical device to be realised. This problem was outlined at the very beginning of work on bubble domain devices by Bonyhard *et al.* (1970) who described the operating region of a given device by a discrete area on a plane having the bias field as one axis and the in-plane drive field as the other. Within this area all the functions involved in the device— straight propagation, corners, turns, generation, transfer gates, etc., would operate. Outside this operating area, first one and then another of the required functions would fail. The same kind of operating margin area in the (B_0, B_D) plane was used by Archer *et al.* (1972) and Kontera *et al.* (1972) to show the effect of varying the overlay bar dimensions and spacing. An experimental study of optimum spacing by Chen and Nelson (1972) also presented the results in this way. The reduction in the operating margins of B_0 and B_D with increasing frequency was shown by a number of authors, in particular by Bosch *et al.* (1973a, 1973b), Parzefall and Littwin (1973), Chang *et al.* (1972) and by Fischer (1971).

All these results suggest that there are two overlay tracks which are superior in performance and these are the **X** bar track of figure 5.15(d) and the double sided track of figure 5.17. The reason for the superiority of these two tracks is that their action involves the simple alternation of the direction of magnetisation along the long direction of an overlay bar while the **T** bar, **Y** bar and chevron patterns call for complicated changes in the magnetic domain structure of the bars during propagation. The **X** bar pattern still requires some complications where the bars cross one another but, never-theless, Parzefall and Littwin (1973) found that this pattern could be oper-ated up to 600 KHz compared to only 200 KHz for a **Y** bar pattern on the same material. Fischer (1971) showed that his double-sided bar track would operate at 700 KHz compared to 200 KHz for a **T** bar pattern on the same material.

The conclusion to be drawn is that the optimum propagation track for a bubble domain device will be a double-sided track, in that a magnetic pattern should be formed upon *both* sides of the bubble domain layer, and that this track should be one which produces a travelling magnetic well of the kind discussed in Chapter 4, § 4.3.5. This travelling magnetic well should move along a path which does not pass completely underneath the permalloy overlay, as it does in the **T** bar, **Y** bar and chevron, so avoiding the formation of complex magnetic domain patterns in the overlay which will introduce serious time delays.

If this conclusion is correct, an important step in bubble domain device technology has been made with the very recent development of sputtered ferrimagnetic films, which support bubble domains, by Chaudari *et al.* (1973). These materials, based upon the alloys of gadolinium, cobalt and iron, were not mentioned in Chapter 3 because they are not yet established as certain contenders in bubble domain device applications. The reason for mentioning them here is that this development shows the way we should go if good propagation tracks are to be made, that is, away from liquid phase epitaxy which requires a perfect single crystal substrate and towards a bubble domain material which can be deposited upon any kind of sub-strate and can, therefore, be deposited *on top of a propagation pattern*. One side of a double-sided propagation track is thus prepared, covered by a few microns of silica spacer, then the bubble domain material is deposited, then more silica spacer and, finally, the other side of the propagation track. The structure of figure 5.17 is an example.

A further point of speculation for the future concerns the size of the overlay pattern needed for bubble domain propagation, relative to the size of the bubbles themselves. In all the patterns shown in figure 5.16, and in the double-sided track of figure 5.17, we find that the width of the overlay bar is about half the bubble domain diameter.

There are very good reasons for expecting the technical limit to the fabrication of such overlay patterns to be found near overlay bar widths of about 0·3 μm. Such patterns can be made by using electron beam techniques (Hu *et al.* 1973). This would set a lower limit to the bubble domain size at about 0·6 μm. Smaller bubbles could be used if a propagation pattern could be developed which would have coarser dimensions than the bubble domains and an example of such a pattern is shown in figure 5.18. This

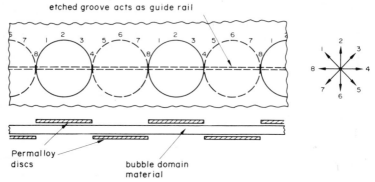

Figure 5.18. The double sided track using permalloy discs. The advantage is that the bubble domain diameter is much smaller than the diameter of the discs.

pattern is due to Bobeck *et al.* (1971b) and uses circular discs of permalloy instead of bars and is a double-sided track so that good performance would be expected. The track works by the same principle as Rossol and Thiele's (1970) well known mobility experiment, described in Chapter 4, § 4.2.2. As shown in figure 4.1, the bubble domain is about one quarter the diameter of the permalloy discs and propagates around their circumference as the in-plane field rotates, alternately transferring its allegiance from a disc at the top of bubble domain material to one at the bottom. This transfer is ensured by some guide rail which runs through the diameters of all the discs. A 'U' turn in such a track is simply made by terminating the pattern and the guide rail.

If the disc diameter was also limited to 0·3 μm by the fabrication process, we would now be able to propagate magnetic bubbles with diameters below 0·1 μm. Figure 3.5 shows that the kind of material which supports such small bubbles is going to have a very high value of $\mu_0 M$ and this will introduce very considerable technical difficulties. If these problems can be overcome, however, a bit density considerably above anything hoped for with semiconducting devices, which are limited by the same problem of making fine line widths in metallisation, will be achieved (Chang 1973).

5.3   Bubble Domain Detection

We shall now consider the problem of bubble domain detection. This is the output end of the device where the problem is to find out if a bubble domain is present at some point within a device or, perhaps, just to detect a bubble domain when it passes through a given point in a device.

A number of detection techniques have been suggested and these were discussed briefly in Chapter 1, § 1.4.4. The discussion here will be concerned with the magneto-resistive detector only, because this detector has proved to be the most successful and belongs in this chapter with magnetic overlay patterns, it being a magnetic overlay itself.

5.3.1   *Magneto-resistance*

The phenomenon of magneto-resistance has been known for a very long time (Bozorth 1961) and, in the majority of materials, is typified by an increase in resistance when the material is magnetised in the same direction as the current flow. Magnetisation in a direction at right angles to the current usually produces a drop in resistance.

In the case of thin films of permalloy, prepared by evaporation, the difference between the resistance of a rectangular element when it is magnetised across the current flow, as shown in figure 5.19(a), and when it is magnetised parallel to the current flow, as shown in figure 5.19(b), will be a few percent of its normal resistance provided the two states shown are near to magnetic

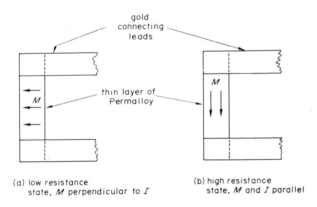

(a) low resistance
    state, *M* perpendicular to *I*

(b) high resistance
    state, *M* and *I* parallel

Figure 5.19. The low and high resistance states of a thin film permalloy sensor element.

saturation. If the magneto-resistive element is not near magnetic saturation, so that the true difference between the situations shown in figures 5.19(a) and (b) is merely a small change in the domain pattern within the permalloy, the change in resistance will be very small indeed.

5.3.2   *Magneto-resistive bubble domain sensors*

Referring back to figure 5.13, it may be seen that the stray field of a bubble domain, when it is located at one end of an overlay bar by means of an in-plane drive field B_D, produces a stray field at the surface of the garnet layer which aids the field B_D on the left hand side of the bubble and opposes it on the right hand side. Figure 5.13 shows that the stray field of the bubble is, in fact, strong enough to cancel out the in-plane field, B_D, on the right hand side and it is this effect which may be used to actuate a magneto-resistive sensor element.

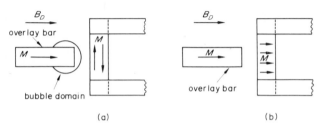

Figure 5.20. The magnetoresistive sensor element is placed where the stray field of the bubble domain cancels out the in-plane field. Shape anisotropy then ensures that the sensor is magnetised along its long dimension. When there is no bubble at the end of the bar (b), the in-plane field saturates the very thin permalloy layer so that its magnetisation is perpendicular to the direction of current flow.

This is shown in figure 5.20(a). The sensor element is placed at the end of an **I** bar in a propagation track so that it is located in the region where the stray bubble field and the in-plane field cancel one another out. Because of its shape anisotropy, the magnetic domain pattern in this thin film permalloy sensor will take up the form shown in figure 5.5; one side will be magnetised to saturation parallel to the current flow while the other side will be magnetised anti-parallel to the current flow. Magneto-resistance is an even effect so that this demagnetised state has the same resistance as the state in which the bar is fully saturated in the same direction as the current.

In figure 5.20(b) we show the situation which applies in the absence of a bubble domain. The sensor element will now be exposed to the full intensity of the in-plane field B_D and become very nearly saturated in a direction perpendicular to the current flow. This is the low resistance state.

The sensor element will only behave in this way if it is an extremely thin layer, much thinner than the overlay bars which were discussed in § 1 above. The sensor described here is due to Almasi *et al.* (1971, 1972) and was described in more detail by Bosch *et al.* (1973a) as shown in figure 5.21. In these designs the permalloy for the sensor was only 0·02 to 0·025 μm thickness compared to 0·25 μm for the overlay.

This difference in permalloy thickness may mean that a separate step in the device fabrication process will be needed to produce the required sensor

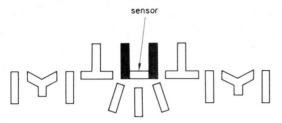

Figure 5.21. The magnetoresistive sensor in a propagation track. The three overlay bars below it stretch the bubble domain out to produce a stronger interaction with the sensor.

elements. This can be avoided if the sensor element can be made from permalloy of the same thickness as the propagation track and this approach is recommended by Bobeck *et al.* (1973) who formed a sensor element from a number of interconnected chevron elements. This structure stretches the bubble domain out during sensing, as the three bars shown in figure 5.21 do, and may increase the coupling between the bubble domain and the sensor.

5.3.3 *Using the magneto-resistive sensor*

Almasi *et al.* (1972) discusses the problem of the voltage which will be induced in the sensor elements on a bubble domain device by the rotating in-plane field. In practice, this is overcome by using two sensors, in close proximity, connected to a differential amplifier as shown in figure 5.22. One

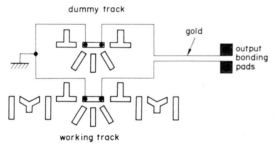

Figure 5.22. A sensor in a dummy track is used to balance out the inductive coupling between a sensor and the rotating drive field.

sensor is associated with a dummy propagation track and produces the magneto-resistive signal from a continuous stream of 'zeros' together with the signal induced by the rotating field. The other sensor produces the magneto-resistive signal from the 'ones' and 'zeros', propagating along the lower track in figure 5.22, and the same induced signal. The amplifier only sees the difference between these two signals which is, consequently, a reading of the binary signal propagating in the lower track.

The equivalent circuit for this arrangement is shown in figure 5.23. Two constant current sources, I, of direct current, supply the two sensors, R_1 and R_2. The input impedance of the amplifier is R_{in}. If we assume that the two sensors are well matched, $R_1 = R_2 = R$ and only one is changing its

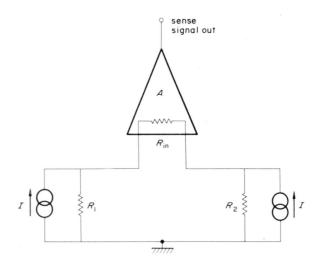

Figure 5.23. The two sensors of figure 5.22, R_1 and R_2, are connected to the input of a differential amplifier, input impedance R_{in}.

resistance in a pulse like manner from R to $R + \Delta R$. The signal voltage to the amplifier is simply

$$v_{in} = (IR)(\Delta R/R)\left(\frac{R_{in}}{2R + R_{in}}\right) \qquad (5.29)$$

which is clearly a maximum when R_{in} is made as large as possible. In bubble domain devices, however, we are working at quite high frequencies and it is useful to consider the condition of maximum power transfer from the sensor. The power input to the differential amplifier is simply v_{in}^2/R_{in}, for a rectangular pulse waveform, and this is a maximum when $R_{in} = 2R$ for this particular connection. The available power from the sensor is then

$$P_{MAX} = \frac{I^2}{8}\left(\frac{\Delta R}{R}\right)^2 R \qquad (5.30)$$

Taking the figures given by Almasi *et al.* (1972) for a typical sensor, $I = 5\,\text{mA}$, $\Delta R/R = 0{\cdot}018$, $R = 75\,\Omega$, we find that the available power is $7{\cdot}5 \times 10^{-9}$ Watts.

It is interesting to note that the magneto-resistive sensor is acting as a power amplifier in this situation because its input power, in a signal sense, is coming from the stray field of the bubble domain which is being detected.

We can obtain a feeling for this power gain by taking the energy of the bubble domain to be of the order of $\mu_0 M^2 h^3$ and divide this by the time taken to sense the bubble domain to obtain the power input. This time is of the order of a quarter of the period of the rotating field. On this basis the power gain of the sensor is

$$G \approx \frac{I^2 (\Delta R/R)^2 R}{32 \mu_0 M^2 h^3 f} \tag{5.31}$$

which comes to ≈ 4 when we substitute the values given by Almasi *et al.* (1972), $\mu_0 M \approx 2 \times 10^{-2}$ T, $h \approx 5 \times 10^{-6}$ m, $f \approx 100$ KHz. This is an interesting result in that the power gain exceeds unity even though we have taken the total energy of the bubble domain as the input energy. In fact, the magneto-resistive sensor is coupled to only a fraction of this energy.

Equation (5.31) shows that the power gain increases as $I^2 R$, the dc power dissipated in the sensor element itself. This is, of course, limited in practice and it is possible to increase the gain by a factor of four by pulsing the current I so that it flows for only one quarter of the rotating field period when a bubble domain is expected. The current is also limited by the fact that the magnetic field produced by the sensor should be very much smaller than the field which is moving the domain, but this limit is far greater than the limit set by the maximum permissible dissipation.

5.3.4 *Other form of bubble domain sensing*

The Hall effect has been used with some success for sensing bubble domains (Strauss and Smith 1970) and optical techniques have been discussed by Almasi (1971). These alternatives are more complex than the magneto-resistive detector and have not found any application in working devices so far.

5.4 Conclusions

In this chapter we have concentrated upon the magnetostatics of single overlay bars and their interaction with magnetic bubble domains. These theoretical problems form the foundation for the design and optimisation of bubble domain propagation tracks and for bubble domain detectors, but there is a great deal more to the bubble domain device than this. The problems associated with transfer gates, bubble–bubble interactions for logical operations between tracks, of the kind described in Chapter 1, § 4.4.4, rely for their solution upon the same theoretical foundation, but call for considerable ingenuity and inventiveness, not only to realise such functions, but also to realise them in such a way that they can be fabricated simply and operate within one device in the same applied fields as all the other operations which must occur at the same time.

A few papers have been published which deal with this problem of realising a complete bubble domain device and these show that this is a field of very rapid technical progress. It is also clear that the usual compromises must be made to overcome practical problems of defects in the materials and difficulties in the fabrication process. These papers, Bosch *et al.* (1973), Bonyhard and Guesic (1973) and Michaelis (1973), to mention the most recent open publications, remind us that the fabrication of a bubble domain device may involve a very large number of difficult steps. For example, to list a typical history, there could be—

(i)   Growth of substrate material.
(ii) Slicing and polishing of substrate.
(iii) LPE growth of the garnet layer.
(iv) Silica layer.
(v)   Permalloy at ≈ 200 Å for sensors.
(vi) Metallisation for sensor connections and control.
(vii) Permalloy at $\approx 0.5\ \mu$ for tracks.
(viii) Construct drive coils and package.

The steps (v), (vi) and (vii) all involve some kind of etching and photolithography, certainly three masks would be needed to deliniate the sensors, metallisation and propagation tracks. However a single etching step can serve to generate the final composite structure.

Compared with some of the techniques which are used at the present time in semiconductor work, which involve diffusion processes and deep etching to the same fine scales which are required for the bubble domain device, the process detailed above appears fairly straightforward. When the metallurgical and crystal growth problems of the bubble domain device have the same background of knowledge which is available for semiconductor device work, and it is very likely that bubble domain devices will be made which involve well over an order of magnitude more data per unit area than the semiconductor devices of that future time (Chang 1973).

References

Almasi, G. S., 1971, *IEEE Trans. on Magnetics*, MAG 7, 370.
Almasi, G. S., Keefe, G. E., Lin, Y. S., and Thompson, D. A., 1971, *J. appl. Phys.*, **42**, 1268.
Almasi, G. S., Keefe, G. E., and Terlep, K. D., 1972, High speed sensing of small bubble domains. *18th Conf. on Mag. and Magnetic Mats.*, Denver, Nov. 1972 (*AIP Conf. Proc.* No. 10, p. 207).
Archer, J. L., Tocci, L. R., George, P. K., and Chen, T. T., 1972, *IEEE Trans. on Magnetics*, MAG 8, 695.
Bobeck, A. H., Fischer, R. F., Perneski, A. J., Remeika, J. P., and van Uitert, L. G., 1969, *IEEE Trans. on Magnetics*, MAG 5, 544.
Bobeck, A. H., Fischer, R. F., and Smith, J. L., 1971a, *AIP Conf. Proc.*, No. 5, p. 45.
Bobeck, A. H., Della Torre, E., Perneski, A. J., and Scovil, H. E. D., 1971(b), U.S. Pat. No. 1241235, *Improvements in or relating to magnetic single wall domain devices*, filed, Aug. 4th. 1971.
Bobeck, A. H., Danylchuk, I., Rossol, F. C., and Strauss, W., 1973, *IEEE Trans. on Magnetics*, MAG **9**, 474.

Bonyhard, P. I., Danylchuk, I., Kish, D. E., and Smith, J. L., 1970, *IEEE Trans. on Magnetics*, MAG 6, 447.

Bonyhard, P. I., and Geusic, J. E., 1973, Paper No. 21.7, *IEEE Trans. on Magnetics*, MAG 9, 433.

Bosch, L. J., Downing, R. A., Keefe, G. E., Rosier, L. L., and Terlap, K. D., 1973a, IBM Research Report, RC 4272, March 16th 1973.

Bosch, L. J., Downing, R. A., and Rosier, L. L., 1973b, *IEEE Trans. on Magnetics*, MAG 9, 481.

Bozorth, R. M., 1961, *Ferromagnetism*. Van Nostrand Publishing Co., Princeton, N.J., p. 745.

Chang, H., Fox, J., Lu, D., and Rosier, L. L., 1972, *IEEE Transactions on Magnetics*, MAG 8, 214.

Chang, H., 1973, Paper No. 4.3, *INTERMAG Conf.*, Washington D.C., April, 1973.

Chaudhari, P., Cuomo, R. J., and Gambino, R. J., 1973, *IBM J. Res. and Dev.*, 17, 66.

Chen, Y. S., and Nelson, T. J., 1972, *IEEE Trans. on Magnetics*, MAG 8, 754.

Copeland, J. A., 1972, *J. appl. Phys.*, 43, 1905.

Danylchuk, I., 1971, *J. appl. Phys.*, 42, 1358.

De Blois, R. W., 1968, *J. appl. Phys.*, 39, 442.

Druvesteyn, W. F., Tjaden, D. L. A., and Dorleijn, W. F., 1972, *Philips Res. Repts.*, 27, 7.

Fischer, R. F., 1971, *IEEE Trans. on Magnetics*, MAG 7, 741.

Gemperle, R., 1966, *Phys. stat. sol.*, 14, 121.

George, P. K., and Chen, T. T., 1972, *Appl. Phys. Lett.*, 21, 263.

George, P. K., and Archer, J. L., 1973a, *J. appl. Phys.*, 44, 444.

George, P. K., and Archer, J. L., 1973b, Paper No. 13.7, *INTERMAG Conf.*, Washington D.C., April, 1973.

Goldstein, R. M., and Copeland, J. A., 1971, *J. appl. Phys.*, 42, 2361.

Hayashi, N., Romankiw, L. T., Chang, H., and Krongelb, S., 1972a, *IEEE Trans. on Magnetics*, MAG 8, 370.

Hayashi, N., Chang, H., Romankiw, L. T., and Krongelb, S., 1972b, *IEEE Trans. on Magnetics*, MAG 7, 16.

Hu, H. L., Hatzakis, M., Giess, E. A., and Plaskett, T. S., 1973, Paper No. 26.5, *INTERMAG Conf.*, Washington, April, 1973.

Jahnke, E., and Emde, F., 1945, *Tables of Functions*. Dover Publications, New York, pp. 73–85.

Johnson, W. A., North, J. C., and Wolfe, R., 1973, Paper No. 21.2, *INTERMAG Conf.*, Washington D.C., April, 1973.

Jones, M. E., 1970, *Ferromagnetism in highly perfect metallic platelets*, Ph.D. thesis, Imperial College, London.

Jones, M. E., 1973, private communications, 3rd April, 1973 and 19th June, 1973.

Kempter, K., 1972, *IEEE Trans. on Magnetics*, MAG 8, 375.

Kotera, Y., Kinoshita, R., Namikata, T., and Nishimura, Y., 1972, *IEEE Trans. on Magnetics*, MAG 8, 673.

Le Craw, R. C., Byrnes, P. A., Johnson, W. A., Levinstein, H. J., Nielsen, J. W., Spiwak, R. R., and Wolfe, R., 1973, *IEEE Trans. on Magnetics*, MAG 9, 422.

Lin, Y. S., 1972, *IEEE Trans. on Magnetics*, MAG 8, 375.

Michaelis, P. C., 1973, *IEEE Trans. on Magnetics*, MAG 9, 436.

Parzefall, F., and Littwin, B., 1973, *IEEE Trans. on Magnetics*, MAG 9, 293.

Rossol, F. C., and Thiele, A. A., 1970, *J. appl. Phys.*, 41, 1163.

Rossol, F. C., 1971, *IEEE Trans. on Magnetics*, MAG 7, 142.

Smythe, W. R., 1968, *Static and dynamic electricity*. McGraw Hill Publishing Co. Inc., New York, N.Y., p. 290.

Sommerfeld, A., 1950, *Mechanics of deformable bodies*. Academic Press, New York, N.Y., p. 147.

Sommerfeld, A., 1952, *Electrodynamics*. Academic Press, New York, N.Y., p. 78.

Strauss, W., and Smith, G. E., 1970, *J. appl. Phys.*, 41, 1169.

Wolfe, R., North, J. C., Johnson, W. A., Spiwak, R. R., Varnerin, L. J., and Fischer, R. F., 1972, Ion implanted patterns for magnetic bubble propagation, *18th Conf. on Mag. and Magnetic Mat.*, Denver, Nov. 1972 (*AIP Conf. Proc. No. 10*, p. 339).

Yamauchi, F., Yoshimi, K., Fujiwara, S., and Furuoya, T., 1972, *IEEE Trans. on Magnetics*, MAG 8, 372.

6 Conclusions

6.1 Bubble Domain Devices

A recent review by Almasi (1973), of the technical problems associated with bubble domain devices, shows that this field is extremely active and full of the most interesting problems. It is hoped that the theory given in the preceding chapters will prove useful to anyone taking up this exciting new work. Here, we have directed attention towards those applications which exploit the intrinsic binary nature of bubble domains and their use in data processing systems. Several papers should be mentioned in this connection which deal with the organisation and mathematics of systems which make use of entities having the interactive properties of bubble domains. These are Beausoleil *et al.* (1972), on organisation, Graham (1970), Sandfort and Burke (1971), Ahamed (1972) and Kluge (1972), on fundamental mathematical theory.

In conclusion, it is interesting to look at some quite different possibilities for the application of bubble domains.

6.2 Display Devices

One of the most interesting aspects of experimental work with magnetic bubble domains is that we can observe these domains during motion by means of the Faraday effect. Even if we have to work with a non-transparent medium, like the metal films mentioned in Chapter 5, § 5.2.4, it is still possible to see the domains by using the polar Kerr effect. An excellent review paper on this and other magneto-optic effects has been given by Palik (1967) and an extensive bibliography by Palik and Henvis (1967). The polar Kerr effect was used by Fowler *et al.* (1963) to display domains in opaque samples of orthoferrite, cut so that the magnetisation was normal to the surface of the sample, as it is in a bubble domain layer, and was also used by Chaudhari *et al.* (1973a, 1973b) in their study of the domain structure and bubble domains in thin films of gadolinium iron and gadolinium cobalt.

With these optical techniques being used by experimental workers for bubble domains, it was natural that a number of suggestions were made for display devices. Shoji (1972) suggested a shift register for counting which produced a numerical display of the number counted by using bubble domains. Kita *et al.* (1972) described a method of shifting two dimensional arrays of bubble domains by means of a two dimensional magnetic overlay pattern. This device could store a two dimensional pattern, which was generated by an electrical input signal, and then shift it either horizontally or vertically. In this way the pattern could be rotated, all these operations being performed by means of electrical signals to the device.

Perhaps the most remarkable work on bubble domain display devices is that published by Ashkin and Dziedzic (1972) who showed that bubble domains may be manipulated in a layer of bubble domain material, having no overlay of any kind, by means of a spot of intense laser light. This effect comes about because of the local temperature gradient which is produced by the laser light. A temperature gradient will produce a gradient in the magnetisation of the material and the force on a bubble domain, in a uniform bias field, is usually such that it will move towards the region of higher magnetisation (Thiele *et al.* 1971) so that we would expect the bubble domain to move away from the laser light spot because the magnetisation usually falls with increasing temperature. In a material like holmium iron garnet, however, (figure 3.2) the opposite could apply and the bubble would move into the heated zone. Thiele *et al.* (1971) also showed that the temperature dependence of the domain wall energy could produce a force upon the bubble domain, when there was a temperature gradient in the material, and that this force usually opposed the force produced by the magnetisation gradient. Boxall (1973) has recently measured the force on a bubble domain in a well defined temperature gradient, using an epitaxial layer of $(EuY)_3(GaFe)_5O_{12}$ and has obtained good agreement with the theory published by Thiele *et al.* (1971).

Ashkin and Dziedzic (1972) also demonstrated the manipulation of stripe domains with laser light and their paper has the most beautiful photographs showing these results. These authors were even able to write their signatures with this technique in a script with characters only 100 μm or so high.

6.3 Bubble Domain Arrays

The bubble domain array has already been introduced here by means of figure 1.9 which shows a fairly well ordered hexagonal array of bubble domains generated by means of a repeated pulsed magnetic field. In figure 1.9, the bubble domains are held in equilibrium with an applied bias field, but it turns out that it is possible to have an hexagonal array of bubble domains in zero applied field. This was first discussed by Nemchik (1969) who demonstrated that these stable bubble domain arrays could exist in thin slices of gadolinium iron garnet. Measurements of the nearest neighbour spacing and domain diameter, in crystals of known thickness, were used to characterise materials by Charap and Nemchik (1969) and the response of bubble domain arrays to pulsed magnetic fields was used to study dynamic properties by Nemchik and Charap (1971). The theory of these stable bubble domain arrays was dealt with in great detail by Druyvesteyn and Dorleijn (1971) and by Cape and Lehman (1971). These authors showed that the bubble domain array was only stable in zero bias field when the material had sufficient wall energy to produce a small contraction in the bubble domain diameter. In other words, the magnetic field produced by all the bubbles surrounding any given bubble in an array is insufficient to stabilise it against the expanding force of its own magnetic field. A further force is required and this comes from the contracting force due to the wall energy, as discussed in Chapter 2, § 2.3.

This results in an interesting magnetisation characteristic for bubble domain arrays which is shown schematically in figure 6.1. If we were to observe the bubble domain lattice, as a Faraday rotation micrograph, under conditions of varying bias field, B_0, we would observe two possible remanent states, R_1 and R_2 in figure 6.1. If we suppose that the magnetisation we

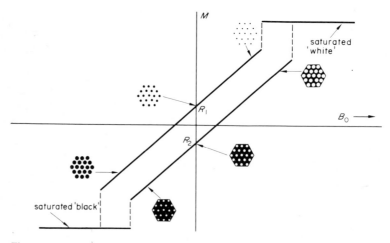

Figure 6.1. A schematic magnetisation characteristic for a bubble domain array. Inset are diagrams showing the domain pattern under various conditions of magnetisation and applied field. The white regions are magnetised in the positive sense, the black regions negative.

choose to call positive appears white in the micrograph, while that which appears black is negative, it is clear that there will be a positive remanent state which will appear as black bubble domains upon a white background. This is the positive remanent state, R_1, because we know that the wall energy causes the bubble domains to contract slightly leaving a volume of material in excess of 50 per cent which is magnetised towards us and appears white. It is then possible to apply an external field, B_0, and run up and down the upper part of the characteristic shown in figure 6.1, contracting and expanding the bubble domains. Too large a positive field causes bubble collapse and positive saturation of the sample. Too large a negative field causes the black bubble domains to run into one another and cause negative saturation.

The other possible remanent state is R_2; white bubbles on a black background. This is a negative remanent state because, again, the wall energy causes the bubbles to contract below the diameter which would cause them to occupy 50 per cent of the volume.

The separation between the two characteristics has been exaggerated in figure 6.1—Cape and Lehman (1971) show that the remanence can only be a few percent of the saturation magnetisation even when $\lambda/h \approx 1$. It is interesting to note that the value of B_0 at which the characteristic cuts the B_0 axis is not a coercive force in any sense. We have here a truly off-set magnetisation characteristic.

Referring again to figure 6.1, it was found (O'Dell 1973) that it was possible to switch discontinuously between the two remanent states R_1 and R_2. The process by which this may occur is shown in figure 6.2. Beginning with state R_2 of figure 6.1, shown as stable state 1 in figure 6.2, a positive field, B_0, is applied which expands the white bubble domains so that they eventually run into one another to form the lattice of small black bubble domains shown in the centre of figure 6.2. The pulsed field is then removed and this lattice of small bubbles, which is magnetostatically unstable, relaxes back to the remanent state R_1. The process shown in figure 6.2 is speculative, although it seems certain that something of this kind must occur because

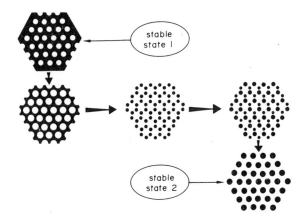

Figure 6.2. The process which may occur during switching between the two remanent states of a bubble domain array (O'Dell, 1973).

it is possible to switch between the two stable states continuously. The process occurs very rapidly, a 0·3 μS pulse being required for bubble domain arrays in a 10 μm thick film of $(EuY)_3(GaFe)_5O_{12}$. It should be possible to verify that the process shown in figure 6.2 really does occur by using the very high speed photographic techniques used by Malozemoff (1973) in his study of bubble domain dynamics.

A further aspect of bubble domain arrays which is of interest is the study of their dynamic behaviour when excited by small oscillating fields. This has been looked at theoretically by Kaczér and Tomáš (1972) and by Höfelt (1973a, 1973b). Work of this kind may be technically important because bubble domain arrays can be made to cover quite large areas of material and excitation of one region of the array with an oscillating field would produce a delayed response in a more distant region. Such an effect would be useful for signal delay applications.

The final point of interest in bubble domain arrays is their property of acting as two-dimensional diffraction gratings. The bubble domain array may be so well ordered and regular that an excellent diffraction pattern is obtained when plane polarised light is passed through the array and then through an analyser, to convert the phase grating properties of the bubble domain array into an amplitude grating and to remove the bright central spot from the diffraction pattern. Photographs of these spot diffraction patterns have been published by Chaudhari *et al.* (1973c) and by Papworth (1973). The ring diffraction patterns which may be produced by the serpentine domains in bubble domain layers, shown here in figure 1.5, have been published by Woolhouse and Chaudhari (1973).

6.4. Conclusion

From the pure scientist's point of view, the magnetic bubble domain has opened up a new field in magnetism for theoretical and experimental work. The highlight here is perhaps the dynamic work reviewed in Chapter 4, particularly when the atomic scale dynamic processes are considered. The bubble domain has produced a great interest in domain wall structure and the problem of understanding the stability and creation of this particularly well defined situation where we have one closed wall surrounding a magnetic domain in a thin sheet of material. In this connection the ring domain or 'hollow bubble' (de Jonge *et al.* 1971) should be mentioned because this unusual domain has perhaps an even greater sensitivity to its wall structure than the simple bubble domain and the problem of its initial creation, and how this effects the wall structure, might be easier to examine experimentally.

From the engineer's point of view the bubble domain device looks full of interest for digital applications, and for the display or pattern manipulation devices which have been reviewed in this last chapter. For digital data processing systems, the power consumption and speed potential of systems using magnetic bubbles (Keyes 1971) look very interesting, particularly when we remember the high packing density which may be possible and the close association of memory and logic which the bubble domain allows.

References

Ahamed, S. V., 1972, *Bell Syst. Tech. J.*, **51**, 461.
Almasi, G. E., 1973, *Proc. IEEE*, **61**, 438.

Ashkin, A., and Dziedzic, J. M., 1972, *Appl. Phys. Lett.*, **21**, 253.

Beausoleil, W. F., Brown, D. T., and Phelps, B. E., 1972, *IBM J. Res. Dev.*, **16**, 587.

Boxall, B. A., 1973, Pers. Comm., 16, Feb. 1973.

Cape, J. A., and Lehman, G. W., 1971, *J. App. Phys.*, **42**, 5732.

Charap, S. H., and Nemchik, J. M., 1969, *IEEE Transactions on Magnetics*, MAG 5, 566.

Chaudhari, P., Cuomo, J. J., and Gambino, R. J., 1973a, *Appl. Phys. Lett.*, **22**, 337.

Chaudhari, P., Cuomo, J. J., and Gambino, R. J., 1973b, *IBM J. Res. Dev.*, **17**, 66.

Chaudhari, P., Cuomo, J. J., and Gambino, R. J., 1973c, *INTERMAG Conf.*, Washington D.C., April, 1973, Paper 17.8.

De Jonge, F. A., Druyvesteyn, W. F., and Verhulst, A. G. H., 1971, *J. Appl. Phys.*, **42**, 1270.

Druyvesteyn, W. F., and Dorleijn, J. W. F., 1971, *Philips Res. Repts.*, **26**, 11.

Fowler, C. A., Fryer, E. M., Brandt, B. L., and Isaacson, R. A., 1963, *J. Appl. Phys.*, **34**, 2064.

Graham, R. L., 1970, *Bell Syst. Tech. J.*, **49**, 1627.

Höfelt, M. H. N., 1973a, *J. Appl. Phys.*, **44**, 414.

Höfelt, M. H. N., 1973b, *INTERMAG Conf.*, Washington D.C., April, 1973, Paper No. 29.6.

Kaczér, J., and Tomáš, I., 1972, *phys. stat. sol. (a)*, **10**, 619.

Keyes, R. W., 1971, *Proc. IEEE*, **59**, 1528.

Kita, Y., Inose, F., and Kasai, M., 1972, *IEEE Trans. on Magnetics*, MAG 8, 367.

Kluge, W., 1971, *Electronics Letters*, **7**, 749.

Malozemoff, A. P., 1973, *IBM Technical Disclosure Bulletin*, **15**, 2756.

Nemchik, J. M., 1969, *J. Appl. Phys.*, **40**, 1086.

Nemchik, J. M., and Charap, S. H., 1971, *Metallurgical Transactions*, **2**, 635.

O'Dell, T. H., 1973, *Phil. Mag.*, **27**, 595.

Palik, E. D., 1967, *Applied Optics*, **6**, 597.

Palik, E. D., and Henvis, B. W., 1967, *Applied Optics*, **6**, 603.

Papworth, K. R., 1974, *Phys. stat. Sol. (a)*, **22**, 373.

Sandfort, R. M., and Burke, E. R., 1971, *IEEE Trans. on Magnetics*, MAG 7, 358.

Shoji, M., 1972, *IEEE Trans. on Magnetics*, MAG 8, 240.

Thiele, A. A., Bobeck, A. H., Della Torre, E., and Gianola, U. F., 1971, *Bell Syst. Tech. J.*, **50**, 711.

Woolhouse, G. R., and Chaudhari, P., 1973, *Phys. stat. Sol. (a)*, **19**, K3.

Subject Index

Index of Principal Authors

158